Number 22 *Energy Policy Review* (£1.75)

Number 23 *Tidal power barrages in the Severn Estuary* (£1.50)

Number 24 *Advisory Council on Energy Conservation: Paper 6: Freight transport: short and medium term considerations* (£1.50)

Number 25 *Advisory Council on Energy Conservation: Paper 7: Report of the Working Group on Buildings* (£1.75)

Number 26 *Advisory Council on Energy Conservation: Paper 8: Energy for Transport* (£1.25)

Number 27 *Severn Barrage Seminar September 7 1977* (£3.25)

Number 28 *Report on research and development 1976–77* (£1.75)

Number 29 *Energy forecasting methodology* (£3.25)

Number 30 *Gas gathering pipeline systems in the North Sea* (£2.50)

Number 31 *Advisory Council on Energy Conservation: Report to the Secretary of State for Energy* (£1.75)

Number 32 *Energy conservation research, development and demonstration* (£2.25)

Number 33 *Energy conservation: scope for new measures and long-term strategy* (£2.00)

Number 34 *Heat loads in British cities* (£2.75)

Number 35 *Combined heat and electrical power generation in the United Kingdom* (£3.75)

Number 36 *Advisory Council on Energy Conservation: Paper 9: Civil aviation: energy considerations* (£2.25)

Number 37 *Advisory Council on Energy Conservation: Paper 10: Report of the Publicity and Education Working Group* (£2)

Number 38 *Report on research and development 1977–78* (£2.75)

Number 39 *Energy technologies for the UK*

Number 40 *Advisory Council on Energy Conservation: report to the Secretary of State for Energy* (£3.25)

Number 41 *National energy policy* (£1.50)

Preface

With the completion of the first phase of the UK's Wave Energy R and D Programme the Wave Energy Steering Committee felt that publication of a review would be useful. Hence this document. It reviews the under-lying technology and describes the status of the UK programme as it stood at November 1978. The detailed results from this programme were given at a Conference held at the Heathrow Hotel, London Airport, on 22nd and 23rd November and the Proceedings will be published separately.

Wave energy is at that state, so familiar in technological innovation, where problems loom larger than solutions. These problems stem from the nature of the resource itself:

—it is diffuse;
—the movements of the sea are slow and irregular;
—the waves arrive at our shore from directions that cover a wide angle.

The diffuseness of the resource is perhaps surprising. Everyone knows how powerful the sea is; the 70–90kW/m of wave front (as it is at Station India, a weathership some 400 miles West of the Hebrides) seem at first sight to reinforce the idea of its power. Actually the energy offered per unit area to any wave power device at Station India or closer inshore would still be 10 or more times less than the energy crossing the primary heat exchanger surfaces of a conventional power station, and the amount actually converted to useful power much less again. This diffuseness implies that large areas of device have to face the sea and, if a nationally appreciable amount of power is to be generated, over long stretches of coastline. The Figures in Chapter 3 illustrate how big are the first generation of machine designs. Their sheer size involves steel and concrete whose structural costs would amount to perhaps 50–60 per cent of the total costs, estimated at £4000–9000 per kW installed capacity (to be compared with £500–1000 for more conventional fossil or nuclear stations).

The second problem is the slow irregular movement of the sea, in complete contrast to the fast regular movements required of machinery if it is to produce electricity efficiently. The lesson from the results so far appears to be that this presents an especially difficult problem for wave power if a mechanical linkage is used at the prime mover stage of conversion; this is because of the difficulties of engineering large mechanical components to take the large forces, and the associated problems of corrosion, reliability and maintenance. Devices based on linking water turbines or air turbines directly to the sea show more promise from these viewpoints; but these are the devices which suffer most from the size and structural cost problems discussed in the previous paragraph. Furthermore, machinery that has to be rated to accommodate seas at the more energetic end of the spectrum will spend much of its time operating well below its rated capacity, because the sea movements vary so widely from minute to minute, day to day and season to season.

The third problem, that of directionality, arises because the devices have to be arranged in a long line, which can face only one direction. Hence the amount of energy from other directions that is available per metre of machine front is reduced according to a simple cosine law.

Putting all these problems together, the estimated cost of electricity from the first generation of wave energy machines is 20–50 p/kWh, compared with around 2 p/kWh annual average from the central electricity generating system.

All those engaged on the programme see the exposure of these harsh realities as the first step towards a clear indication of priorities in the further work to be done. Indeed only a few months after the November conference, teams were already claiming significant reductions in cost arising both from evolutionary development in conceptual designs as well as from completely new ideas. One new idea in particular shows promise of significant cost reduction though its engineering feasibility needs to be proved (the Flexible Bag, pages 31 *et seq*.).

The message is that this report can provide only an interim statement on the potential and prospects for wave energy. Much more knowledge has to be obtained before firm conclusions can be drawn.

F. J. P. Clarke
Chairman,
Wave Energy Steering Committee

February 1979

Note: The subject of wave energy is moving rapidly. This report represents the situation at the end of 1978.

Contents

Page		
viii		List of figures
1		*Over-view*
4	1	*Introduction*
4		Historical development of the UK programme
6		Programmes in other countries
6		Organisation and aims of the UK programme
8	2	*Wave data*
8		Wave physics
15		Wave data
17		Wave data programme
17		Wave data and the design of converters
19		Wave data and resource assessment
21	3	*Engineering development of converters*
22		Four main converter designs
22		Rectifier
24		Oscillating Water Column
27		Raft
28		Duck
29		Alternative concepts
35	4	*General aspects of structural design*
35		Evaluation of wave-induced motions and loads
37		Evaluation of structural response
38		Structural materials
41	5	*Mooring*
41		Mooring systems
43		Mooring components
44		An approach to the mooring problems
47		Concluding remarks
48	6	*Energy conversion and transmission*
50		Conversion and transmission systems
52		Primary power take-off
56		Conversion and transmission
65		Future studies
66	7	*Electricity supply system aspects of wave energy*
66		System characteristics of wave energy
72		System implications of the characteristics
73		Economic assessment of wave energy
76		Future role of wave energy

Page		
80	8	*Environmental and social aspects*
82		Wave climate and the shore line
84		Possible interactions with fishery activities
85		Navigation of ships
87		Economic and social development of local communities
89		Concluding remarks

Appendices

90	1	Principal contractors in the Department of Energy programme (at September 1978)
91	2	Wave Energy Steering Committee (1978)
92	3	Technical Advisory Groups
95	4	Main assumptions used in calculating breakeven capital costs of wave energy

List of figures

Page	Figure	
6	1.1	The organisation of the Department of Energy R and D programme on wave energy
8	2.1	An ideal wave
9	2.2	Waves on a beach
10	2.3	Trochoidal wave
11	2.4	Particle motion
12	2.5	Typical wave record
13	2.6	Spectrum derived from the record of Fig 2.5
14	2.7	Pierson-Moskowitz spectrum for a wind speed of 10 m/s
14	2.8	Scatter diagram from South Uist during the year 1976–77
14	2.9	Fraction of time that the power in the sea exceeds a given level at OWS India during winter months
14	2.10	Double peaked spectrum from data collected off South Uist
15	2.11	'K-space' diagram for the spectrum of Fig 2.10
15	2.12	Wave crests corresponding to the wave vectors of Fig 2.11
16	2.13	Wave-data collecting stations of particular relevance to the wave energy programme
17	2.14	Distribution of power by frequency for the whole year together with predicted efficiency curves for a particular design of duck converter of 6, 10 and 16 m diameter
19	2.15	Some possible locations for wave energy devices
f.p. 22	3.1	Artist's impression of one form of the Rectifier
f.p. 23	3.2	Artist's impression of one form of the Oscillating Water Column Converter
25	3.3	Operating principle of one form of the Oscillating Water Column Converter, as developed at the National Engineering Laboratory
26	3.4	Outline shapes for an Oscillating Water Column Converter, used in Japanese model tests
f.p. 26	3.5	Artist's impression of one form of Raft Converter
f.p. 27	3.6	Artist's impression of one form of the Duck
30	3.7	Classification of converters by three geometrical properties
31	3.8	The Duct (a new concept by Vickers Ltd)
32	3.9	Vickers device working cycle
33	3.10	Illustration of the working cycle of the flexible bag
48	6.1	Derivation of power spectrum from a typical spectrum of wave heights
49	6.2	Typical efficiency variation for a model tested in a narrow laboratory tank
49	6.3	Effect of peak equipment rating on the delivered power
50	6.4	Interaction of the conversion characteristic and the power extracted by the device
50	6.5	Basic flow diagram
51	6.6	Conceptual conversion and transmission systems
53	6.7	Two examples of air turbine characteristics
55	6.8	Interaction of inertia and turbine characteristics

Page	Figure	
56	6.9	Variation of electrical power output
58	6.10	Preferred electrical scheme
59	6.11	Major routes of the 400/275 kV transmission system in Great Britain
60	6.12	Hydraulic power transmission
62	6.13	Schematic flow diagram for hydrogen production
63	6.14	Possible arrangements of chemical energy carrier systems
67	7.1	Mean monthly wave energy supply
68	7.2	Calculated output from a wave energy device
69	7.3	Wave energy contributing to electricity supply
70	7.4	Effect on variability of averaging the coincident outputs from two devices about 250 km apart in mid-winter
71	7.5	Effect on load factor of reducing equipment rating
75	7.6	Target capital costs for wave energy to be economic
77	7.7	An illustration of the possible roles for wave energy
81	8.1	Principal area studied in the preliminary review: between the 10–50 km lines off the Outer Hebrides
83	8.2	Schematic view of a converter array
87	8.3	Fishing activities in the vicinity of the Moray Firth
88	8.4	Systems decisions needed in order to carry out a detailed examination of the amenity consequences of wave energy for the Outer Hebrides

Waves

Over-view

Part of the immense solar energy input to the earth is converted by natural processes into energy associated with ocean waves. The geographical location of the United Kingdom renders it one of the world's most favoured countries with respect to the potential availability of wave energy. In principle, the waves reaching our coastal waters from the North Atlantic might satisfy a considerable fraction of our electricity demand provided that reasonably high overall conversion efficiencies can be achieved.

Inventors have recognised the power of the sea for many decades, and there has been no lack of ideas on how it might be tapped. But none of the ideas was developed on a substantial scale, since ample and relatively cheap supplies of other resources were always available. In recent years, however, there has been a growing recognition that—on a world scale—the presently used forms of energy may become too expensive, too scarce or otherwise unavailable to meet our energy needs by themselves. The Government's responsibility is to ensure that as wide a range as possible of energy supply options are available when they may be needed. Research and development can provide the necessary technical and economic data on which the ultimate choices can be made.

Within this context, the Government announced in 1976 the start of an R and D programme on wave energy for which the first phase was to be a feasibility study lasting for two years. The funding level has been increased twice since that time to maintain the momentum of the programme in the light of technical progress.

The programme has had three main components:

—exploratory development of several different engineering concepts of wave energy converter;

—supporting research in relevant engineering and scientific areas:
- the collection and analysis of wave data,
- analysis of the structural response to wave-induced motions,
- mooring,
- energy conversion and transmission,
- environmental aspects;

—working up preliminary reference designs of full scale stations for technical and economic appraisal.

The purpose of this paper is to review the present state of knowledge of wave energy in the light of the achievements of the first two-year phase of the programme.

Development of the converters

Four potential designs of converter were adopted for initial study, since the sparse data available were insufficient to enable a single concept to be chosen with confidence. Proposals for alternative concepts are received on a continuing basis and are assessed against a number of criteria: two of them have so far been added to the programme in order to explore new principles.

Apart from the basic technical differences the six designs differ from each other in their degree of complexity and their state of development, as described in Chapter 3. Work on two of the designs has been advanced from laboratory wave tanks to the testing of 1/10th scale models in natural open water at Loch Ness and in the Solent. For all the designs, a combination of theoretical studies, laboratory work and engineering appraisal has clarified the factors which will prove to be the most crucial in determining which of them could be chosen for more extensive development.

The programme has progressed from establishing the scientific feasibility of wave energy converters to confirming the engineering feasibility of designing and building some of the designs. In very broad terms:

- the early part of the programme placed considerable emphasis on optimising the efficiency of extraction of the wave energy and proving the scientific feasibility;
- the present stage is concerned with the technical viability and is identifying the main cost centres in the designs, which can then be tackled by further R and D;
- the immediate future must also place emphasis not only on the problems of construction, operation and maintenance, and on ways in which unit costs may be reduced, but also on the ability to survive in the most severe wave conditions.

Whilst the technical feasibility of some types of converter has been established, we are far from the stage of recommending that a full scale generating station should be built. Of the four original concepts, no single design has yet emerged which is outstandingly better than the other designs when all factors are taken into account.

The designs have changed considerably in the course of the feasibility study, and a continuing process of evolution can be expected as in the early stages of any technical development programme. The optimum design may emerge from further changes in one of the original concepts, from a synthesis of ideas or from an alternative concept. The wave energy is distributed over a wide frequency and energy bandwidth and no design has yet been optimised to operate at or near peak efficiency over the whole spectrum. However, one of the new concepts introduced into the programme recently may offer significant advantages in this respect.

Supporting research

The extent of the available data on waves in the sea areas of primary interest is inadequate as yet for the full assessment of the resource. A start has been made in collecting and analysing new data, which will take several years to reach a satisfactory level. The results so far confirm the general point that the locations around the United Kingdom with the most abundant wave energy lie to the west of the Outer Hebrides, where several hundred km of searoom are available with average annual power levels in the range 35–60 kW/m of wavefront.

The majority of the designs under consideration are free-floating and the converters would operate on or near the water surface—one of the most hostile environments for engineering structures. The ultimate feasibility, technical and economic, of all designs of floating converter will depend upon extensive further work on mooring and anchoring. Whilst over-designed mooring systems based on present knowledge have allowed the open water trials to proceed with the objective of gaining experience, the existing knowledge is not adequate to design cost-effective mooring systems which will ensure survival at full scale under storm conditions.

Considerable progress has been made in assessing and understanding suitable energy conversion and transmission systems for the various designs of converter, but much more remains to be done to arrive at the most cost-effective solutions. The general engineering difficulties are quite basic and are related to the properties of the natural wave spectrum:

- the conversion system must be able to handle large short-term variations in the instantaneous power level;
- the peak power level in the sea (of the order of 10,000 kW/m) can be many times greater than the average power level (a few tens of kW/m);
- the primary output is not in a form which can be handled conventionally (it is, of course, variable with time in a complex way).

Moreover, apart from the randomness, other general problems arise from the low energy density of the input and the relatively low speeds and frequency of movement induced by the waves. Engineering devices to transmit large amounts of energy under such conditions must themselves be large, heavy and expensive. The efficient generation of electricity requires machinery operating at relatively high and preferably constant speed. The transition from the one regime to the other appears to be more straightforward for systems involving air turbines than for those which do not: some designs of converter may prove to be intractable in this respect.

Many possible forms of energy transmission to the mainland have been reviewed, as summarised in Chapter 6. Whilst it has been recommended that several options should be kept open in the continuing studies it is likely that most attention will be given to electricity. The overall flow of energy from the waves to a final user connected to the electricity grid involves many separate steps, each of which can involve loss of some of the energy. This can have a considerable influence on the system economics and further work in this area will need to concentrate on both reducing the number of steps and increasing the efficiency (including the directional efficiency of the converters themselves) of those which must remain. Unless this can be achieved the usable resource will be only a small proportion of our needs: some pointers to substantial improvements are beginning to emerge.

Environmental studies have not revealed any major detrimental effects of the converters provided they are well offshore. More information is needed on the behaviour of salmon and herring off the Outer Hebrides to confirm that the fisheries would not be affected significantly by the widespread installation of converters.

Concluding remarks

The costing studies of the reference designs which have been evolved so far indicate that wave-produced electricity is likely to be expensive compared with either nuclear or fossil fuels unless some major breakthrough in the engineering can be achieved. However, this does not imply that the possibility of wave energy should be abandoned at this stage. It must be emphasised that the subject is still at a very early state of development and many unknown factors remain to be resolved. Under these circumstances, wave energy is best regarded at present as a possible insurance technology—the consequences of failure of one of our existing major energy supplies are so severe that it is worth paying an insurance premium to explore fully the alternatives. Nevertheless, the evidence from the feasibility study so far does not allow a recommendation for a full-scale development programme to be made at this time. Much more can be achieved to explore and then to narrow the design choices by continuing work at about the 1/10th scale coupled with, on the one hand, limited trials of some critical components at larger scale and, on the other hand, further creative engineering on the drawing board and laboratory work in a new generation of wave tanks (of which the forerunner has been successfully commissioned at Edinburgh University).

The programme has generated a broad basis of knowledge of all aspects of wave energy which did not exist before, so that we can now identify clearly the critical problems to be tackled by further work.

1 Introduction

Wave energy is a derivative of the solar energy input to the earth, which is accumulated on open water surfaces by the action of the winds. The world-wide wave power potential can be related to the distribution of winds: the strongest winds are located approximately between latitudes 40° and 60° in both the northern and southern hemispheres, with a small number of subsidiary localised areas at about 10°. The equatorial regions are comparatively free of wind and the polar regions are either ice bound or continental. The eastern sides of oceans in the main wind belts, being the downward ends of the fetches, have the highest concentrations of wave energy. It will be seen, therefore, that the UK, geographically located on the eastern side of the Atlantic Ocean and in the right latitude, is one of the world's most favoured countries with respect to the potential availability of wave energy.

Although the existing wave data on a world basis are sparse the UK, being a maritime nation, has rather more data available than elsewhere. Although, as will be shown in Chapter 2, the amount of detail is as yet insufficient for the final design and installation of wave powered generating stations, nevertheless substantial visual and some instrumental observations (for example from Ocean Weather Ship 'India' stationed at 59°N, 19°W in the North Atlantic, see Fig 2.13) over a number of years made it possible to arrive at a general figure for the amount of wave energy which may be available. The data suggested in the early 1970s that the average wave energy to the north-west of the UK could be about 70 MW/km of wave frontage. Since, in principle, several hundred km frontage of generating stations would be possible, it appeared that wave energy, if converted to electricity, could satisfy a considerable fraction of our electricity demand.

Furthermore, in the idealised situation of a constant wind speed of, say, 20 knots, and a 50 per cent efficient extraction device, the locally generated wave energy beneath the winds could be extracted about every few tens of km. Theoretically, therefore, one would not be limited to a single line of extraction devices, but could envisage several relatively parallel lines extending at intervals well out into the ocean. The design of the stations more distant from the shoreline would almost certainly have to differ from those closer inshore: the possibility has not been pursued as yet. The situation is in fact much more complex, depending upon the importance attached to the swell component of the waves (see Chapter 2), but this might present an increasing resource in the long-term future if the early development work is successful.

1.1 Historical development of the UK programme

Waves are irregular in size and frequency. Moreover the surface of the sea is one of the most hostile environments for engineering structures and materials. However, there has been no lack of ideas on how to recover wave energy for useful purposes. Several hundred patents were granted in the UK over the past hundred years or so relating to wave powered generators.

Government interest in the UK in wave energy began formally in 1974 with the publication of a report entitled *Energy conservation* by the Central Policy Review Staff.* This report started from the recognition of the vulnerability of the UK's energy intensive economy to disruptions in supply which had been illustrated vividly by the Middle East crisis. In reviewing the possibility of developing new contributions to electricity supply from inexhaustible supplies of energy, the report highlighted wave energy

*HMSO, 1974

and recommended that the first stage of a full technical and economic appraisal should be put in hand. A detailed introductory assessment of the large scale generation of electricity from ocean waves was initiated in February 1974 at the National Engineering Laboratory (NEL) by the newly formed Department of Energy. Based on the results of this review the Government announced in April 1976 the start of a two-year study costing about £1M, the aim of which was to establish the feasibility of the large scale extraction of power from sea waves and to generate information which would enable the cost of further development to be estimated.

The properties of ocean waves which are likely to be most relevant to the extraction of energy are:

—variations in the slope and height of the travelling waves;
—subsurface pressure variations;
—subsurface fluid particle motion.

The careful classification of energy extraction concepts in the NEL assessment led to the choice of the first of these properties as the main basis for the Government-funded study. Whilst this continues to be the case, some attention has been given to the possible attractions of using the other properties and work on two promising concepts is described later.

At the start of the study the information available was insufficient to allow a firm choice of a single engineering concept to be made from the wide variety which was possible. Consequently it was decided that four basic designs should be investigated, utilising radically different concepts for the primary conversion of the wave energy to mechanical energy. It was anticipated that as the work proceeded it would be possible to narrow the choice progressively until the optimum system remained for final development. Such a final choice has not yet been made.

The four designs which formed the major part of the two-year programme are:

—oscillating vanes ('ducks') invented by S H Salter at Edinburgh University and studied further there and by Sea Energy Associates Ltd using effort based at Lanchester Polytechnic;
—wave contouring rafts invented by Sir Christopher Cockerell and studied by Wavepower Ltd;
—oscillating water columns studied at the National Engineering Laboratory;
—wave rectifiers invented by R C H Russell and studied at the Hydraulics Research Station.

These are described in Chapter 3.

The engineering research on the converters has been supported by general research in several problem areas common to all of them, for example:

—the collection and analysis of wave data;
—studies of the effect of wave forces on the structures of the devices;
—anchoring and mooring problems;
—power generation and transmission to shore;
—the environmental effects.

This work is described in Chapters 2 and 4–8, and a list of the main contractors is shown in Appendix 1.

At the end of the first year, by April 1977, the progress had been sufficiently encouraging for the Government to decide that the programme should be increased from a funding level of £1M to £2.5M. This would allow further work on the collection and analysis of wave data, on the effect of the waves on the structures of the converters, and further work on forms of transmission of the energy to shore, an extension of the engineering testing to the 1/10th scale in natural waves in open water, and the construction of a new wave tank at Edinburgh University in which mixed sea conditions could be simulated for more relevant small-scale experiments. An additional £2.9M was allocated in June 1978 primarily in order to extend the development work at 1/10th scale in open water but also to allow some preliminary development and testing of vital components at a larger scale. The state of knowledge was not yet sufficient to allow development and demonstration of complete larger scale converters to proceed, or to narrow the choice with confidence. The optimum design may not be one of those on which the programme was based originally, but will evolve from an amalgam of ideas through interactions which are already taking place within the programme and from the introduction of completely new concepts as noted in Chapter 3. It is intended that the programme will be reviewed annually from the spring of 1979.

1.2 Programmes in other countries

Considerable progress has been made in Japan over about the same period of time. There the development effort has concentrated on the concept of an oscillating water column for the transfer of energy from the waves to air as the working fluid to drive a turbine. Pioneering work by Masuda has led to the installation in Japanese waters of more than 300 generators in the 70–120 watt region of output for powering light-buoys and lighthouses. A study of large wavepower generators began at the Japan Marine Science and Technology Centre in 1974, based upon Masuda's ideas. After two years of basic study and work with small models in water tanks, a large scale device was developed in 1976, named Kaimei. Its maximum output was expected to be about 2 MW in a sea state characterised by a wave height of 3 metres. Preliminary trials up to a power level of 0.6 MW in the Sea of Japan took place in the autumn of 1978. These are being followed by more extensive trials in which the UK is participating through the International Energy Agency.

A Working Party on Ocean Energy Systems of the International Energy Agency was formed in March 1976 in order to bring together those member countries of the OECD which had a common interest in developing a range of ocean energy sources. This led to the formal signing of an agreement for a collaborative programme on wave energy in May 1978, the participants being Japan, USA, UK and Canada, joined subsequently by Eire. Some further details of this programme are given in Chapter 3. There is also active interest in wave energy in Norway.

1.3 Organisation and aims of the UK programme

The organisation of the Department of Energy programme is shown in Figure 1.1. It is managed by the Energy Technology Support Unit on behalf of the Department of Energy, with advice from the Wave Energy Steering Committee (WESC), the composition of which is shown in Appendix 2. Broad guidance and advice on the recommendations of the committee are provided by ACORD (the Advisory Council on Research and Development for Fuel and Power) which is under the chairmanship of the Chief Scientist. The work of the commit-

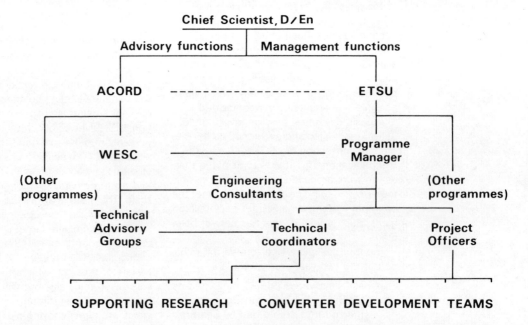

Fig 1.1 *The organisation of the Department of Energy R and D programme on wave energy*

ACORD—Advisory Council on Research and Development for Fuel and Power
ETSU—Energy Technology Support Unit
WESC—Wave Energy Steering Committee

tee and of the converter development teams has been supported in the general research areas through a series of Technical Advisory Groups (TAGS) (see Appendix 3).

Some basic research on novel forms of wave energy converter is supported by the Science Research Council, which is represented on the Wave Energy Steering Committee. If they continue to show promise at the stage when the Council's support becomes no longer appropriate, further exploration can be incorporated as part of the main wave power programme.

This paper is an updated survey of the state of the art on wave energy technology, based upon the very considerable progress which has been made since the original NEL survey in 1974. It is essentially a progress report on the original two-year feasibility study set up by the Department of Energy.

The main achievement of the two-year programme has been to progress from establishing the *theoretical* feasibility of wave power, by very small scale experimental work in laboratory wave tanks, to a first examination of the *engineering* feasibility and cost including experience of larger scale models in a scaled but real sea environment. In general terms, the aims of the original £1M programme have been met, but there is still much to be achieved before wave energy could become an engineering reality.

Although the main elements of the programme have been towards the possibility of developing a significant contribution to our electricity supply from wave energy, this is not the only way in which wave energy could be used. It is possible in principle, for instance, that the output could be used locally to produce essential, energy intensive materials such as fertilisers. These possibilities are being kept under review. Whatever output is finally decided on it is quite evident that, large though the basic resource may be, wave power will not be won easily. The work so far has highlighted a need to push many aspects of engineering beyond the existing state of knowledge into novel operating regimes.

Within the context of the overall strategy outlined in the 1978 Green Paper on Energy Policy* the renewable sources of energy, of which wave energy is one, can be regarded as potential *supplements* to our energy supplies or as *insurances* against possible physical shortages of energy in the long-term future (for instance possible major shortfalls in the contributions from nuclear power or coal in the early years of the next century).

The concept of wave energy as a supplementary supply, for instance of electricity, implies that it could be foreseen as becoming competitive with other sources of electricity. As will be seen later in the paper, the predictions so far on the cost of wave-produced electricity are tending to confirm the original views expressed in the CPRS paper in 1974 that it is likely to be considerably more expensive than nuclear power. Whilst an important aim of a continuing research and development programme must be to reduce the predicted costs, wave energy is better regarded as an insurance technology.

Energy policy. a consultative document. HMSO, 1978. Cmnd. 7101

2 Wave data

A basic knowledge of the properties of waves and of the sources of wave data is so fundamental to an appreciation of all aspects of wave energy converters that it is essential for the first of the technical chapters of this report to deal with this subject. The development of the required data and the formulation and prosecution of a relevant research programme are the responsibility of Technical Advisory Group 2 (TAG 2) of the Wave Energy Steering Committee (see Appendix 3). After a brief description of the appropriate aspects of wave physics, this chapter will outline the work which is being supported through TAG 2.

2.1 Wave physics

Waves are disturbances in the water surface. We are interested primarily in progressive waves, which carry energy from one place to another, rather than standing waves which form when waves are reflected from objects. The larger waves carry most energy.

In order to consider the relevant properties, it is useful to start from an ideal wave in deep water. The wave is assumed to be sinusoidal, small amplitude, monochromatic and to have straight wave-fronts (Figure 2.1). The wave speed (or phase speed) is

$$C = \frac{L}{T}. \qquad (2.1)$$

It is possible to obtain some useful relationships from a mathematical consideration of the general properties of fluids. For instance:

$$C^2 = \frac{g}{K}\tanh Kh \qquad (2.2)$$

where K is the wave number $2\pi/L$, h is the water depth and g is the gravitational acceleration (9.81 m/s^2).

If Kh is large, for deep water, tanh Kh = 1, so

$$C^2 \approx \frac{g}{K} = \frac{gL}{2\pi}. \qquad (2.3)$$

This is an important property of deep water waves—*the speed is proportional to the*

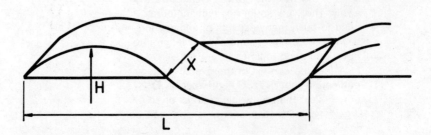

Fig 2.1 *An ideal wave*

square root of the wavelength. Thus the longest waves from a distant storm reach the shore, or a converter, first. The speed of a group of waves, on the other hand, is half the phase speed. Waves start at the back of the group and die out at the front. Knowledge of the speed is vital for wave forecasting.

Combining equation 2.3 with equation 2.1 gives

$$L = \frac{gT^2}{2\pi} \quad (2.4)$$

and

$$C = \frac{gT}{2\pi} \quad (2.5)$$

Thus, for these ideal waves the speed in metres per second is 1.6 times the period in seconds and the wavelength in metres is 1.6 times the square of the period in seconds. It must be stressed that this only approximates the behaviour of real waves.

To make the situation more realistic, one can consider next the effect of shallow water. If the depth is small, Kh is small and tanh Kh≈Kh so that equation 2.2 reduces to

$$C^2 \approx gh. \quad (2.6)$$

Thus the speed is now proportional to the square root of the depth. Waves can be focussed or bent by the process of refraction if the depth is changed. A common example is the fact that waves near a beach are almost always parallel to it (see Fig 2.2). The part of the wavefront in deep water moves faster than that in shallow water and the wavefront swings round.

Waves can also be focussed by undersea hills and sand banks. It has been suggested that artificial undersea plates could focus waves on to wave energy converters or collectors, with the idea of using fewer, larger installations. This suggestion may not have much relevance to the UK wave energy programme because of the large tidal range around the coasts and the absence of suitable estuaries to focus the waves into. In addition, non-linearities may reduce the efficiency of the process. The possibility is, however, being kept under review.

What is meant by the expressions 'deep' and 'shallow'? It is usually accepted that deep water is where the depth is greater than half a wavelength. Shallow water starts where the depth is less than about 1/20 of the wavelength. In between there could be up to about 20 per cent error in the use of the equations 2.3 to 2.6.

The waves considered so far have had relatively small heights compared with their wavelengths. Steep waves are not sinusoidal but have the shape of a rounded tro-

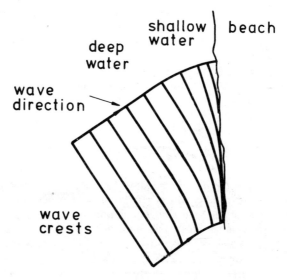

Fig 2.2 *Waves on a beach*

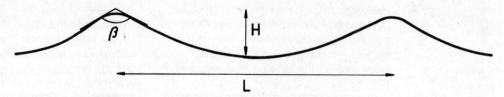

Fig 2.3 *Trochoidal wave*

choid (Fig 2.3). Whilst they are not common in real seas, they are important because they are closer than small waves to the breaking phenomenon.

Such waves have a maximum steepness (H/L) of 1/7. The steepest angle β at the crest is 120°, beyond which the waves start to break. The steepness in deep water is determined largely by the energy input to the sea—high winds raise steep waves which become shallower with time. Large breaking waves have, in the past, caused considerable damage to ships at sea. The forces on structures due to breaking waves are being examined in the wave energy programme: means of minimising them at the converter design stage include the development of converters which operate below the surface or which will submerge in heavy seas.

Energy in a wave

The theoretical derivation of the energy in a wave is a complex matter. The following simplified treatment produces some useful relationships.

In the idealised wave represented in Fig 2.1, the mass of water in the upper half of the wave is

$$\frac{x\rho LH}{2\pi}$$

where ρ is the density of water. The height of the centre of gravity is $\pi H/16$ above the mean level and so the change in potential energy when a crest 'falls into' a trough is

$$\frac{x\rho LH}{2\pi} \cdot g \cdot \frac{2\pi H}{16} = \frac{g\rho LH^2 x}{16}$$

This energy is required to deform the water surface. Kinetic energy also is present because of the wave motion, and in deep water the kinetic energy is equal to the potential energy. Thus the total energy per metre of wave front is

$$\frac{g\rho LH^2}{8}$$

and the power (rate of transfer of energy) is

$$\frac{g\rho LH^2}{16\,T}.$$

since the wave energy travels at the group velocity. Combining this with equation 2.4 gives **power as a function of wave height and wave length**

$$P = \rho\frac{H^2}{16}\sqrt{\left(g^3\,\frac{L}{2\pi}\right)} \quad \text{per metre} \qquad (2.7)$$

or **power as a function of height and period**

$$P = \frac{\rho g^2 H^2}{32} \cdot \frac{T}{\pi} \quad \text{per metre} \qquad (2.8)$$

Energy distribution in a wave

Although a wave travels with a constant speed the water in it does not move at the same speed, or even in the same way as the wave front. In fact the motion of a water particle is roughly circular in deep water and this is why a light object is not pushed along appreciably by small waves. Another important point is that the amount of water movement decreases rapidly with depth and the energy content falls off rapidly as a consequence. The water particles move in circles whose radii are proportional to

$$e^{-2\pi d/L}$$

as shown in Fig 2.4(a). For example, at a depth of half a wavelength the radius is reduced to 1/23 of the radius at the surface.

The fraction f of the wave energy existing between the surface and a depth d is

$$f = 1 - e^{-4\pi d/L} \qquad (2.9)$$

so that, for instance, over 95 per cent is present in the first quarter wavelength down.

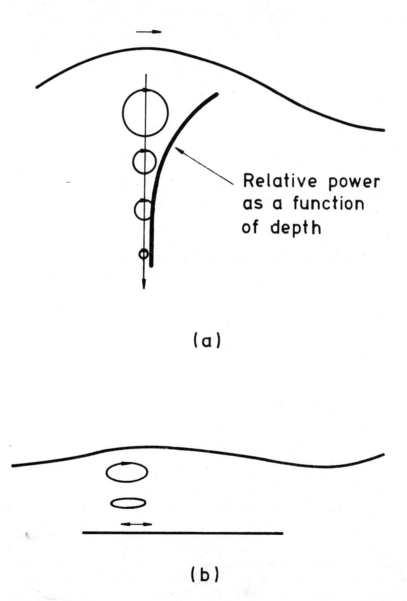

(a)

(b)

Fig 2.4 *Particle motion*
(a) in an ideal deep-water wave
(b) in shallow water

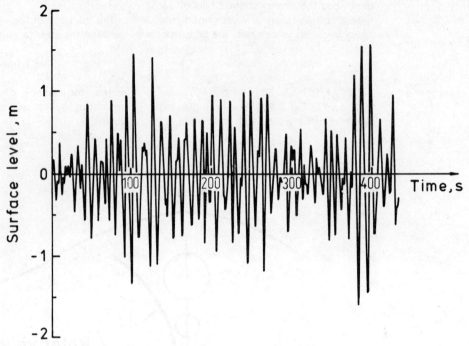

Fig 2.5 *Typical wave record*

The pressure fluctuations in deep water due to a wave passing over obey a similar exponential law to the particle motion, and at a depth of half a wavelength the fluctuations are 1/23 of the pressure fluctuations near the surface. In very shallow water the fluctuations do not decay with depth, and the particle motions become elliptical as shown in Fig 2.4(b).

Despite these relationships, it may be possible in principle for converters to extract significant energy at depths well below the surface, and a relevant design which is being studied is noted in Chapter 3.

Wave momentum
Converters which extract energy from a wave also destroy momentum, and this gives rise to a force on the body of the converter. If F is the average force per unit length on a body when absorbing all the energy in a wave, then the force exerted when the body reflects all the energy is 2F, and the force when it transmits all the energy is zero. Wave energy devices are usually neither perfect absorbers nor perfect reflectors, and there is a net mooring force.

In deep water

$$F = \rho \frac{gH^2}{16} \qquad (2.10)$$

For example, a 3 m wave would, if perfectly absorbed, exert an average force of 5.6 tonnes/m of wavefront on the body absorbing it.

However, waves can also exert other forces than those due to the destruction of momentum. The force due to a head of water produced by a wave can be large and that produced by a breaking wave can be very large. The mooring forces on wave energy converters are likely to be many times larger than those experienced by ships. Ships moored in a storm turn into the prevailing wind (and wave) direction so that the area exposed to the waves is as small as possible. In addition, ships are designed to absorb as little wave energy as possible. The subject is treated in more detail in Chapter 5.

Real waves
The waves described above are idealised models of those which exist in the sea. Real waves are a *mixture* of heights, periods, wavelengths and direction and the

proper analytical use of wave data requires a statistical approach. The usual form that wave data take is a record of the height of the water surface as a function of time at a fixed position in space. Fig 2.5 shows a typical record. From this the height and period can be obtained as follows.

If y_i is the water level at instant i relative to the mean water level then the root mean square elevation

$$\sigma = \sqrt{\frac{\sum_{i=1}^{n} y_i^2}{n}} \quad (2.11)$$

where the average is over n samples of the water height, subject to the proviso that the samples are closer in time than half the highest period present in the record.

A traditional definition closer to the intuitive idea of wave height is that for the 'significant height'. This is the average height of the highest one-third of the waves. Similarly, the 'significant period' is the average period of these particular waves. (The 'one-third' referred to is the third of the total number, not one-third of the height.)

$H_{\frac{1}{3}}$ and $T_{\frac{1}{3}}$ defined in this way are, however, unsatisfactory in practice. Not only are they inconvenient to measure by modern methods, but the definition of 'a wave' causes confusion, and they cannot be related exactly to any other wave parameters in a real sea. Thus, most workers now use a significant wave height parameter H_s defined by

$$H_s = 4\sigma \quad (2.12)$$

and a wave period T_z called the 'zero-crossing period' and defined by

$$T_z = n_z/D \quad (2.13)$$

where n_z is the number of times the water surface moves through its mean level in an upwards direction in a record of duration D seconds.

When there is only a narrow range of frequencies present, $H_s = H_{1/3}$ and $T_z = T_{1/3}$, but this is only approximately true for a real sea.

The power (in kW per metre of wave front) then approximates to

$$P \approx 0.55 \, H_s^2 \, T_z \quad (2.14)$$

(cf. equation 2.8).

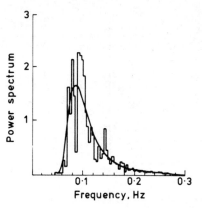

Fig 2.6 *Spectrum derived from the record of Fig. 2.5*

Whilst this can be a useful rule of thumb, calculating the wave energy from H_s and T_z is not a very accurate procedure because the sea frequently contains waves of quite different heights and periods which obviously cannot be represented accurately by one pair of parameters. A record of water level versus time can be turned by the process of Fourier transformation into a spectrum, as in Fig 2.6, where the y coordinate is a measure of energy density and the x coordinate is frequency. This shows that most of the waves have frequencies f of ~0.1 Hz, which corresponds to periods of ~10 seconds.

Theoretical spectra can be calculated and compared with real spectra. The Pierson-Moskowitz spectrum is widely used, and is applicable to a fully developed sea, that is, one where the wind has been blowing in a constant direction for a long enough time for the waves to have stopped growing. Fig 2.7 shows the shape.

With this spectral form it is possible to derive a model wave climate from the wind speed, and the analytic form of the spectrum is useful for predicting the performance of converters in more realistic seas.

The shape of the spectrum is defined by the function

$$S(f) = 5(10^{-4} f^{-5} e^{-(4.2/f^4 W^4)}) \, m^2/Hz \quad (2.15)$$

where W is the wind speed in m/s.

There are many other statistical properties of waves which are useful. If the nth spectral moment M_n is defined as

$$M_n = \int f^n S(f) df \quad (2.16)$$

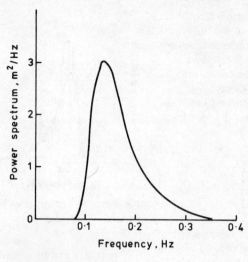

Fig 2.7 *Pierson-Moskowitz spectrum for a wind speed of 10 m/s*

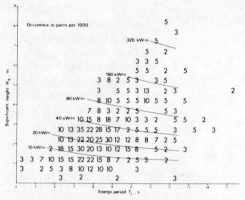

Fig 2.8 *Scatter diagram from South Uist during the year 1976–77*

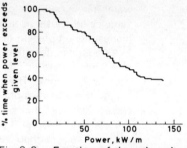

Fig 2.9 *Fraction of time that the power in the sea exceeds a given level at OWS India during winter months*

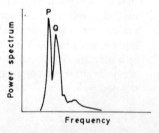

Fig 2.10 *Double peaked spectrum from data collected off South Uist*

the significant wave height

$$H_s = 4\sqrt{M_o} \qquad (2.17)$$

and

$$\sigma = \sqrt{M_o} \qquad (2.18)$$

the mean zero crossing period

$$T_z = \sqrt{(M_o/M_2)} \qquad (2.19)$$

The energy period

$$T_e = \frac{M_{-1}}{M_o} \qquad (2.20)$$

and for most seas

$$T_e \approx 1.12\, T_z \qquad (2.21)$$

The significance of the spectral moments is that they can readily be calculated by computer using digitised wave data.

Wave data may be presented usefully in other forms. Fig 2.8 shows a scatter diagram for data from South Uist. The numbers on the graph represent the fractional occurrence of each significant wave height and period throughout the winter months (December–February). The contours are lines of constant power level, and from the number of occurrences above and below a given power exceedance diagrams like those of Fig 2.9 can be drawn.

Although the spectra described above are much better approximations than the assumption that sea waves are monochromatic, even they do not describe the sea state accurately in many cases. Natural spectra with two peaks are found frequently (Fig 2.10).

Both 'sea' and 'swell' can be present at the same time. 'Sea' is generated by local winds blowing over fetches up to, say, 100 km and is characterised by relatively short wavelengths, over a large range of frequencies and directions. 'Swell', on the other hand, is generated by more distant storms, where the waves have had a reasonable length of time to develop. Much of it can lie outside the idealised Pierson-Moskowitz spectrum. Its importance lies in the fact that it can carry a substantial fraction of the wave energy at a given point and may arrive from directions different from that of the 'sea'.

Directional properties of waves

It is unfortunate for the development of wave energy converters that at the present time instrumental measurements of wave *direction* are scarce. Most of the studies so far use parametric formulae developed by Mitsuyasu, where a 'spreading function' is applied to a one dimensional spectrum of the type discussed above. The spreading function distributes the energy at each frequency over a range of directions, and a typical function is

$$f(\theta) = \cos^s(\tfrac{1}{2}(\theta - \theta_0)) \qquad (2.22)$$

where

$$s = 15.85 \left(\frac{f}{f_0}\right)^5 \quad \text{for } f \leq f_0$$

$$= 15.85 \left(\frac{f_0}{f}\right)^{2.5} \quad \text{for } f > f_0$$

where θ_0 and f_0 are the principal wave direction and the spectral peak frequency.

This relationship assumes a single peak spectrum. A useful way of visualising directional spectra with an arbitrary number of peaks is a diagram of 'k-space', as follows: vectors of length

$$K = \frac{2\pi}{L}$$

are drawn at angles corresponding to the wave directions, where L is the wavelength of that corresponding wave.

Thus the spectrum with two narrow peaks from Fig 2.10 corresponds to Fig 2.11, where the waves come from directions θ_A and θ_B. A picture of the crests of the waves would appear as in Fig 2.12.

In reality, the spectral peaks will be broad and will correspond to a range of wave directions, so that points P and Q (Fig 2.11) will be spread into fuzzy areas.

Figure 2.12 shows that there is a maximum distance D between two wave crests in a 'mixed' sea. If the directions of the waves are more nearly the same this 'long-crested' distance gets longer, and this has

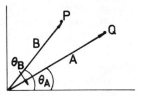

Fig 2.11 *'K-space' diagram for the spectrum of Fig 2.10 (assuming directions for the two principal waves)*

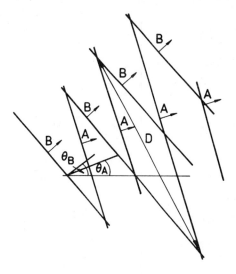

Fig 2.12 *Wave crests corresponding to the wave vectors of Fig 2.11*

a significant bearing on the design of some converters. For example, strings of ducks may use a 'spine' to provide a reference for each duck motion to react against. This works only if the duck string passes through both crests and troughs, so the net spine motion is small. Another problem arises in the extreme case of a spine with its ends on crests and its centre over a trough—there will then be a significant bending moment on the spine. Research into the measurement and prediction of long-crested waves is an important part of the programme.

2.2 Wave data

Most of the historical accumulation of data on waves at sea is based on simple visual observation. The height observed is close to H_s, although perhaps about 20 per cent low, and the period corresponds roughly to T_z. A useful feature of visual observations is that the principal wave direction is usually given. Several million visual wave observations have been recorded.

Shipborne wave recorders are mounted on lightships and weather ships. They provide a record of wave height as a function of time, from which H_s and T_z may be measured. Although the data are of high quality and obtained over a long period of time there are only a few such recorders operating. Many of the data used in the early stages of the wave energy programme had been collected previously at OWS India, situated at 59°N latitude, 19°W longitude.

Waverider buoys use electronic integrators to convert buoy acceleration into displacement, so that a waveheight/time record is obtained. A 27 MHz transmitter sends data back to shore and 20 minutes of data are recorded every few hours. The limited range of the transmitter and the large amount of interference at this frequency limit the location of buoys to less

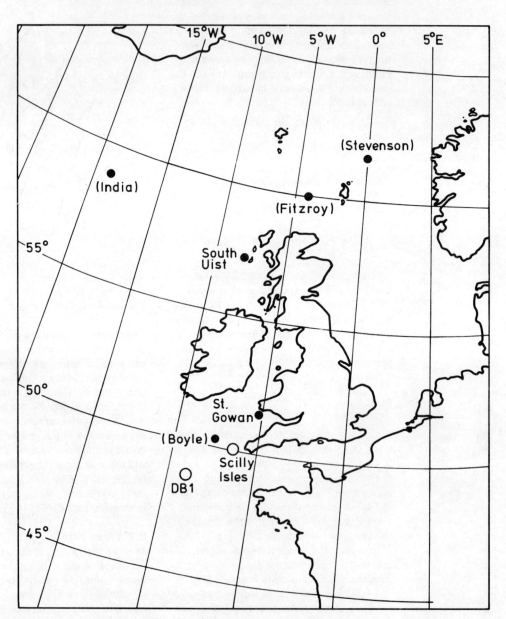

Fig 2.13 *Wave-data collecting stations of particular relevance to the wave energy programme*
( ) denotes no longer operating
O denotes new buoys for which data are not yet available

than about 30 km offshore, but the data obtained from them is very valuable.

Directional measurements can be made with a buoy equipped with several sensors to respond to the different components of acceleration. All the methods so far described only give the state of the sea at a particular location, however. Information about waves over a large area is very scarce. Some airborne photographic and radar observations have been made, but the observations were intended to test the techniques rather than provide a regular data collection service.

The detailed data, necessary both to assess the resource accurately and to design converters, have been collected so far from a relatively small number of sites as shown on Fig 2.13. Wave data are required for many purposes and at the start of the wave energy programme reliance had to be placed entirely on installations which were not ideal for its purpose. Apart from OWS India, data from a buoy off South Uist have become of particular importance to the wave energy programme.

Wave data programme

Examination has shown that the extent of the available wave data is inadequate for detailed wave energy studies. A substantial amount of work has been done with the data from OWS India: although power levels are high there, sites nearer the shore are preferred for wave energy converters. A preliminary inspection of the data from South Uist shows a lower power level based on two years observation, but it is not known how localised this may be. Obviously, more extensive data must be collected from sites closer inshore than OWS India in order to establish the overall size of the resource.

Work in progress in mid-1978 included the following:

- analysis of data from South Uist, Fitzroy, Boyle and St Gowan to produce one dimensional spectra;
- synthesis of directional spectra from South Uist and St Gowan data;
- installation of the Scilly Isles buoy, and overflights using a radar altimeter to check the accuracy of synthesised directional spectra.

Plans are being made also for work on:
- installation of new buoys in locations determined by the specific needs of the wave energy programme;
- making wave predictions and testing them against observations;
- investigations of directional wave probes;
- breaking wave analysis and prediction;
- crest length measurements;
- investigation of the usefulness of satellite radar data.

Wave data and the design of converters

The design of a converter is affected strongly by the wavelengths experienced in the sea and their relative frequencies of occurrence. The efficiency of a converter is usually a strong function of the size of the converter relative to a 'typical' wavelength. Converters with a broad efficiency peak can usually capture more energy from the sea than those which have a sharp efficiency peak. Figure 2.14 shows the power spectrum from OWS India compared with predicted efficiency curves for a particular design of duck converter of 6, 10 and 16 m diameter.

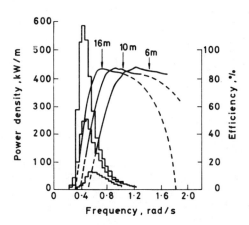

Fig 2.14 *Distribution of power by frequency for the whole year (middle histogram), winter (top) and summer (bottom); together with predicted efficiency curves for a particular design of duck converter of 6, 10 and 16 m diameter (dashed lines indicate extrapolation outside experimental range)*

Although the 6 m duck captures much less power per metre over the whole year than the 16 m duck, a very rough calculation shows that it might require $(6/16)^2 \approx 1/7$ the volume of material to construct and therefore be cheaper. Thus data on the wavelength or frequency range to be experienced by a converter are vital for the cost optimisation of the design. Many designs of converter work by resonant absorption, which implies that response to a particular range of wave periods is important. In this case the mass, stiffness and damping must be adjusted to maximise the capture of energy. Because the sea contains a complex mixture of wave periods, the designer attempts to broaden the resonating peak in order to collect more energy. Although individual swell waves can carry very large amounts of energy, it does not appear practicable to match the converters to their appropriate frequency range—the economic penalty is too great.

There have been several proposals for designs which depend on sub-surface pressure variations. As equation 2.9 shows, in deep water the pressure fluctuations are very small even a short distance below the surface. Such converters have to work in relatively shallow water where the pressure fluctuations are transmitted downwards without a great deal of attenuation. There are fewer such sites, and the energy there is lower.

. Information on the direction of the waves is necessary for orientating the converters. Most linear designs extract energy most efficiently from waves parallel to their long dimension, and the spread of directions, in conjunction with the efficiency curves and the spread of wavelengths, will govern how much energy is lost from this effect. In some cases the nature of the available sites will govern the orientation of the converters and research is needed on directional spectra for these sites. The concept of 'point absorbers'—converters small compared with a wave-length—which can accept energy from a wide range of directions—is important here. Research programmes on the characteristics of point absorbers are being carried out in collaboration with the Central Electricity Generating Board (CEGB) and University departments. One result of this research might be to determine whether linear devices need to be 'fully filled' or whether, for example, a gap of the order of a raft width can be left between rafts in a chain whilst still absorbing energy from the gap. This could have important implications for the economics.

An important area of wave physics with considerable implications for the design of converters is the question of vertical transport of energy. A subsurface or bottom-mounted converter might in principle extract a large fraction of the energy in a wave, implying that energy extracted at one depth is 'replenished' by energy travelling vertically from another depth. Very little engineering research has been done on this problem, largely because existing marine structures are not designed to absorb energy at selected depths. It is possible also that the physical theory may break down when applied to the real sea.

Wave height is important in the design of converters for two reasons. The first is that the power output is proportional to the square of the wave height, so that accurate data on wave heights are required to calculate power levels and forces in mechanical components. The second is a more subtle effect, and is due to non-linearity. Water waves, unlike for example radio waves, are not linear. Two wave groups that pass through each other will not continue undisturbed afterwards unless their amplitudes were very small to begin with. This effect is very difficult to treat theoretically, and accurate representations of real seas in wave tanks are needed to test model converters. Waves become non-linear when their steepness exceeds about 1/25. Although such steep waves are not common in the sea, they might occur, for example, because of reflections from the face of a wave energy device. One result of tests so far carried out is that some designs perform better in large waves than expected, but others perform worse.

The incidence of breaking waves in the open seas is said to be low from visual observations, but data buoys record high, steep waves much more frequently than would be expected. The fact that even large ships can be damaged seriously by breaking waves implies that wave energy converters will have to be designed with large factors of safety. Research on breaking waves is included in the wave energy research programme. One way of reducing the problem might be to use submerged converters, where the conditions are less extreme, but as noted above this may simply imply exchanging one set of problems for a different set.

Wave data and resource assessment

Assessment of the energy available is a vital part of the research programme. However, existing measurements of the wave climate around the United Kingdom are not yet detailed enough to provide more than a first guess at the size of the resource. Due to the random and seasonal nature of winds and waves, measurements have to be taken over many years to produce an accurate base for prediction. Progress in resource assessment has been as follows:

—It was recognised early in the programme that the west coast of the British Isles received higher energy waves than the east, particularly in Scotland, where high wave energy densities occur reasonably close to shore.

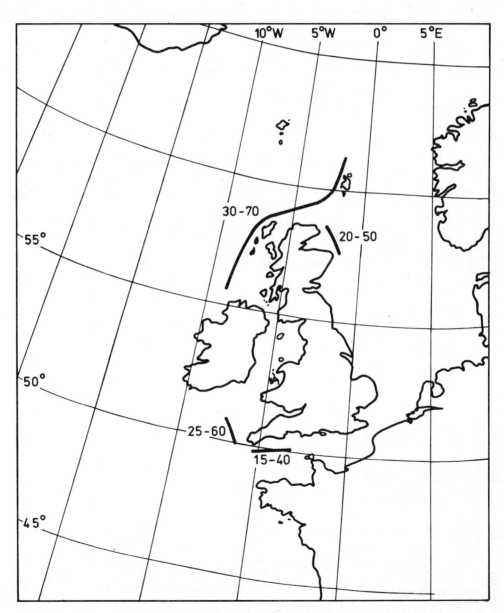

Fig 2.15 *Some possible locations for wave energy devices. An estimate of the energy available is given in kW/m annual average*

— Preliminary studies of wave and wind records indicated possible annual average power levels of about 70 kW/m within a reasonable distance of the Outer Hebrides.
— To check power levels closer to shore, a waverider buoy was installed 11 miles southwest of Benbecula, off South Uist, early in 1976.
— Data from the first two years of operation have been analysed recently, and indicate power levels of about 50 kW/m passing from various directions over a circle of 1 m diameter, which corresponds to about 35 kW/m length of coastline. However, extrapolation of this figure to the whole of the Hebrides area must be regarded with some caution.
— Data from weatherships are being analysed.

Results so far confirm that the best locations for wave energy converters are off the Hebrides where several hundred kilometres of sea room are available at power levels of between 30–60 kW/m annual average. Other possible sites are off the southwestern peninsula, and off the northwest coast of Scotland. Figure 2.15 shows these sites, with an indication of the amount of energy which may be potentially available from them.

3 Engineering development of converters

The analysis by the National Engineering Laboratory of a wide variety of engineering concepts of wave energy converter, which preceded the setting up of the two-year feasibility study as noted in Chapter 1, was hampered by a general lack of data. A firm choice of the best concept could not be made. Four concepts were adopted for the main thrust of the programme, therefore, each to be investigated by a separate team (the Device Teams).

There are many ways of classifying converter designs for analytical and comparative purposes, of which the following basic sub-division by the mode of operation is useful for considering the UK programme:

— Rectifiers
— Tuned oscillators
— Untuned dampers.

There is one example of a rectifier in the programme: the design under investigation at the Hydraulics Research Station. Wave energy is converted into the potential energy of a large mass of water in a reservoir. Since it is fixed to the seabed, the converter has to resist the total momentum of the oncoming waves, which inevitably leads to a relatively massive construction per unit of energy output. On the other hand, the seabed mounting eliminates all the mooring problems which can be serious in some of the free-floating designs.

The other three major designs which have been investigated can all be classed as tuned oscillators, though they aim to solve the engineering problems in very different ways. They operate by providing a 'tuned' working component which is excited by a chosen part of the wave spectrum and damped by a power take-off system. The working component has to present an extensive working surface to the incident sea, for instance a width up to 50 m for an individual raft.

Investigations have been made of a theory that a tuned oscillator could extract energy from many times its own width, but it does not appear possible to develop a practical system of this type: non-linear behaviour invalidates the theory for all but the smallest wave amplitudes and the bandwidth for energy absorption is too narrow.

The body, or reference frame, against which the working component of a tuned oscillator has to react must be relatively unmoving. Either this is accomplished by providing a heavy mass or, in the duck, by a spine which is effective by balancing forces over a wavelength or more. The mass approach leads by definition to heavy structures: the spine may in some circumstances be significantly more economic in the use of materials but has inherently more difficult structural problems.

An example of an untuned damper which has recently been introduced to the programme will be described in the section on alternative concepts. Like the oscillators, there is a need for a large working surface and a 'fixed' frame of reference, though in this case it is possible that the frame can be considerably lighter.

In general, the development of wave energy converters will involve the design and construction of massive engineering structures. The large size is really inherent in the nature of the resource: whilst there is potentially a large amount of energy available it is in a relatively dispersed, or dilute form compared with, say, coal or oil in which natural processes have concentrated the energy content over thousands of years. Whilst there is no prospect in sight of the development of a very small ingenious device which will do the same job as the current conceptual designs of large converters, it is by no means certain that the minimum in specific size and cost per unit output has yet been realised: the fact that this is so provides one of the underlying objectives for an ongoing development programme.

In April 1977 the Wave Energy Steering Committee recognised that expert advice on the problems of large scale engineering design and production was desirable even in the early, formative stages of the development—both to the Committee itself and to the individual Device Teams. Rendel, Palmer & Tritton, in association with Kennedy & Donkin, were appointed as Consulting Engineers and submitted a preliminary report in August 1977. The introduction of the consulting engineers was of great benefit to the programme in evolving reference designs of each converter and in assessing their engineering and cost.

A general point made in the 1977 report was that at that stage the testing of models and the theoretical understanding of wave energy converters were ahead of design development. The predicted costs of the 1977 reference designs were high but, in view of the very early stage of the whole project, not too high to discourage continuation of the work. The main civil engineering structure of the converters was identified as the key cost centre and four specific ways of reducing the overall costs were highlighted:

—reorganisation of the basic design layouts on a more cost effective basis;
—more accurate stressing and proportioning of the designs;
—some radical changes of components within the design formats;
—a more searching approach to the costing of components when designs are firmed up.

In general, it was recommended that the continuing effort would be better directed towards reducing the converter costs than in seeking marginal improvements in overall efficiency.

The Device Teams have begun to tackle these matters during the past year. However, progress throughout the programme is an iterative process and the criteria for optimising the converter designs are still somewhat speculative. As designs evolve so the goals set for them may change in line with what may be achievable. In particular, the value placed on the product—for instance electricity—may change with the predicted firmness (reliability). Cost targets will depend upon the role envisaged for wave-produced electricity, for instance whether it is seen as a supplementary or insurance source of power, as noted in Chapter 1. The size and even the type of converter will depend upon the goals which are set.

The following sections of this chapter provide a brief description of the engineering development so far. An important point is that no one of the original four designs has yet emerged as outstandingly better than the others when all factors (including cost) are taken into account. However, there is now a much more highly developed sense of realism in all the designs and a very much deeper appreciation of the engineering problems which still have to be overcome if wave energy is to become a reliable source of power. The designs have undergone considerable change during the course of the programme. A further phase is necessary to determine whether cost-effective solutions can be evolved to the critical problems which have now been highlighted (and which differ for each design). Those designs for which no credible solution can be evolved will, of course, be eliminated from the programme. The optimum design may yet evolve from one of the original concepts but may also be a synthesis between them through a dynamic interaction of ideas, or develop from one of the new alternatives.

3.1 Four main converter designs

Rectifier

As will be noted in more detail in Chapter 6, the Rectifier offers the possibility of a straightforward solution to the problems of primary power take-off, a water turbine of near-conventional design, and is consequently relatively simple to envisage. It is, therefore, conveniently described first, and is illustrated in Fig 3.1.

A single converter is a large, rectangular, hollow caisson with a system of internal reservoirs and an integral module containing the power plant. The vertical seaward face is provided throughout with an array of panels of one-way flap valves, their hinges aligned vertically, and arranged alternately to allow water to flow in or out.

The converter is divided internally into two reservoirs by a slab which extends

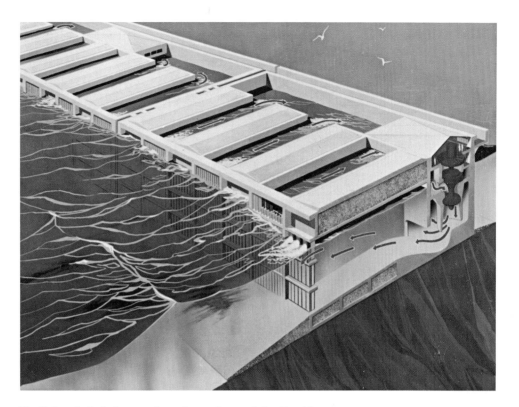

Fig 3.1   *Artist's impression of one form of the Rectifier*

Fig 3.2   *Artist's impression of one form of the Oscillating Water Column Converter*

horizontally for the full length. The reservoir above the slab has a free water surface open to the sky. The outlet reservoir below the slab is provided with a free surface by chambers which project upwards at intervals through the dividing slab and which have a vented roof level with the top of the outer walls of the converter. Two large low-head Kaplan turbines are placed in the flow path between the reservoirs. The generators, driven directly through vertical shafts, are in a machine house above the turbines.

A head difference is maintained between the water in the reservoirs by inlet flap valves to the upper reservoir collecting water during impinging wave crests and outlet flap valves to the lower reservoir discharging water during wave troughs. Flow between the reservoirs drives the turbines.

The caissons would be constructed in special yards and then floated into position and sunk on to the seabed, which would have to be specially prepared beforehand. A line of caissons resting on the seabed in approximately 15 m depth of water would be aligned parallel to the general shore line. The caissons would project above the water level by some 5 m and be sited between 1–5 km from the shore depending on the gradient of the seabed. Discrete lines of converters would be separated by gaps of the order of 1–2 km, depending on contours and the requirements of navigation.

If the width of a converter presented to the oncoming waves is 100 m, the peak generating plant capacity would be 3 MW on the current reference design. Whilst the structure would be a fairly conventional caisson, capable of construction without significant new technical problems, the large size would restrict the number of sites suitable for construction. As an indication of the size, the standard caisson currently envisaged would contain some 56,000 tonnes of concrete.

Progress and outstanding problems. The experimental testing programme has been running at the Hydraulics Research Station for about 18 months and has been based upon 1/30 scale models in monochromatic waves. The concept is basically simple but, as with the other concepts, there are still considerable unsubstantiated features in the reference design.

The concept allows a wide freedom in the organisation of the system of valves, reservoirs and structures. Several significantly different options are available for the basic layout and the problem is to identify the one which is most cost-effective. The 1977 review highlighted the need to reduce the size per unit of output and some progress has been made in this respect, for instance by reducing the front to back dimension and by changing the configuration of the interlocking upper and lower reservoirs. However, this still remains one of the most substantial problems.

Early in the programme it was thought possible that the high throughput at very low head demanded of the water turbines would lead to design requirements to which feasible solutions might not exist. However, subsequent evaluation has indicated technical feasibility: indeed the slow speed Kaplan turbines should be relatively robust and reliable.

The Device Team has developed a tapered rubber flap valve which operates well in the model conditions. Full scale flaps would require reinforcement with metal, and further work on their design is needed.

Whilst fouling by marine organisms may have a detrimental effect on the operation of the flaps and their hinges, there is considerable experience of the behaviour of rubber in sea water. As noted in Chapter 8, a survey by the Consulting Engineers of suitable locations off the Outer Hebrides has revealed that there may be considerable problems of choking of the system by drifting kelp, and more detailed evaluation of this point is needed.

In summary, the testing and design work has reached a stage where the concept is considered to be technically feasible to build but success would be needed in solving problems in the following areas before the concept could be acceptable overall:

—fouling and clogging by marine organisms;
—reduction in size per unit of output (to reduce cost);
—long term reliability of the rectifying valves.

A relatively unattractive feature of the Rectifier which is emerging is that it is not likely to be able to capture as much of the wave energy resource as some of the free-floating designs. A preliminary survey has shown considerable restrictions on the location of the converters off the Outer Hebrides due to misalignment of the seabed contours (which will determine the orientation of the working face) with respect to the most frequent direction of the oncoming waves and to shadow effects from headlands. As noted in Chapter 2, the accumulation of detailed information on the directional properties of the wave spectrum can have an important influence on the design of the converters and indeed on their overall viability.

Oscillating Water Column

The concept of a water column oscillating vertically in tune with the oncoming waves has for many years been thought of as a promising wave energy converter, and has indeed been put to use in Japan on a very small scale of output for powering navigation buoys.

The trapped oscillating column of water can be made to do work in a number of ways, of which the favoured method is to allow it to operate as a piston to pump air through a turbine. Even when this choice has been made there remain a wide variety of configurations and the combination of parameters which will determine the efficiency of the converter is particularly complex. A more detailed discussion of the primary power take-off is given in Chapter 6.

A Device Team at the National Engineering Laboratory has been developing this concept. Their version uses the inertia of a converter and the surrounding water to react the forces on the water column. The converter is envisaged as a floating structure approximately equal in depth and width and with a length about three times the width, see Fig 3.2.

The front face (that facing the oncoming waves) has openings to the sea and contains the oscillating water columns which pump air to drive low pressure turbines and which in turn drive electrical generators which are all housed behind the front chambers in the main body of the converter. The operating principle is illustrated in Fig 3.3

The interaction of the converter with the waves induces vertical, horizontal and rotational motion of the structure as well as oscillations of the column itself and each of these motions causes waves to be generated. The converter also scatters the sea waves because it presents a physical barrier to them and all these radiated wave components combine vectorially to give the resultant radiated wave field. The ideal would be for them all to cancel so that no energy is radiated by the converter and the sole effect is the complete absorption of the incident waves. In the compact free-floating converter such as we are considering here, where the motions of the structure are appreciable, the extent to which cancellation of the wave components can be achieved depends primarily on the shape. Extensive tank testing, mostly at 1/100 scale, has led to the development of shapes for which a high degree of cancellation is achieved in typical seas resulting in a high extraction efficiency. In big seas the converter tends to ride like a cork.

This converter has two very desirable features. Firstly, the large wave forces are at no point concentrated into point loads in the structure or the turbine but remain as distributed pressures and, secondly, the turbine and other machinery is accessible for maintenance and repair.

Progress and outstanding problems. An interim reference design has been developed to the stage where construction in reinforced concrete or structural steelwork is practicable. Progress has been made on the internal layout, valves and plant, and a comprehensive testing programme in wave tanks has been completed. The primary power take-off via the air turbine is a positive asset in that it is reasonable within present technology. Attention has been given in the detail of the design of that part of the structure which encases the water columns to ensure that water cannot enter the air turbines under abnormal conditions. Whilst there are still some problems in this area they are regarded as solvable.

Despite the technical progress which has given confidence in the engineering feasibility, however, the cost-effectiveness of the particular design solutions so far adopted needs to be improved for three main reasons:

— although smaller than some of the waves in which it would have to operate, the structure illustrated in Fig 3.3 is very large;

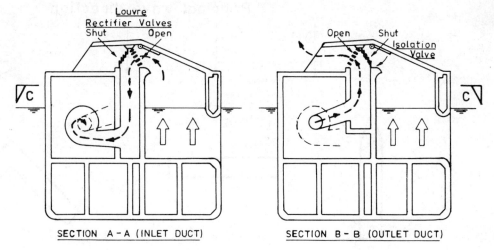

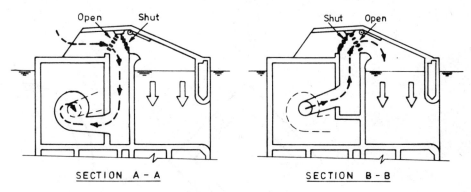

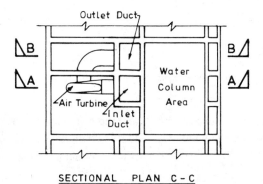

Fig 3.3 *Operating principle of one form of the Oscillating Water Column Converter, as developed at the National Engineering Laboratory*

—the geometry and mass determine that construction could take place only on a limited number of sites of the type used for constructing gravity platforms for the North Sea;

—the shape, alignment and motion of the converter combine to make a particularly severe mooring problem in terms of force and excursion.

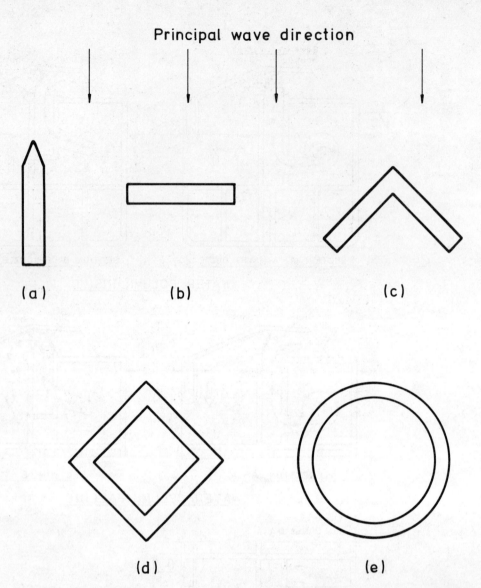

Fig 3.4 *Outline shapes for an Oscillating Water Column Converter, used in Japanese model tests*

It should now be possible to build on the much improved technical understanding of this converter to investigate the many other types of layout in order to move towards the one which is most cost-effective.

International collaboration. A different layout based on the principle of the oscillating water column is being developed by the Japan Marine Science and Technology Centre as noted in Chapter 1.2. Several overall shapes of the converter were tested as small models in wave tanks, Fig 3.4, and the ship-type (a) was chosen for further development for two basic reasons:

—it should have the lowest mooring forces;
—the ease of construction, based upon shipbuilding techniques.

On the other hand this shape is likely to collect energy from a narrower wave front than the other shapes. It will be noted that outline shape (b) is the one which has formed the basis of the work at the National Engineering laboratory in the UK programme: it will collect energy along a

Fig 3.5 *Artist's impression of one form of Raft Converter*

Fig 3.6 *Artist's impression of one form of the Duck*

broader front at the expense of higher mooring forces.

The Japanese programme has moved forward to the point of testing during 1978 a vessel about 3 km off Yura in the Sea of Japan. The vessel, named Kaimei, has provision for 22 separate air pump rooms operating from oscillating water columns which are open to the sea at the bottom (the vessel is kept afloat by buoyancy compartments). Up to 10 electric generators connected to air turbines can be installed eventually, with a possible combined peak output of 2 MW, although not all of the positions will be in use initially.

Testing of models in wave tanks showed that the best efficiency would be obtained when the length of the converter is 1.2–1.7 times the anticipated average wave length. Analysis of the wave data from the Japan Sea then indicated that the overall length of the converter should be about 80 m. The design wave for the Kaimei has a wave length of 75 m (period, 7 sec) and significant height of 4 m, thus the vessel is somewhat smaller than, but approaching, that which would be the design optimum for operation in the North Atlantic.

The United Kingdom, together with Canada and the USA, has accepted an offer by the Japanese to enter into a joint programme of further work with the Kaimei under the auspices of the International Energy Agency. Amongst other clauses, the agreement presents an apportunity to test a UK-built air turbine system on the Kaimei during 1979.*

The possibility of gaining experience of operation in the sea simultaneously with the testing of some of the UK designs in a more highly instrumented manner at the 1/10th scale in more sheltered waters will be an instructive addition to the overall UK programme, and there should be considerable benefits arising from mutual detailed evaluation of the highly complex problems of the design of wave energy converters.

Raft

The concept of the wave-contouring raft, as invented by Sir Christopher Cockerell, is being developed by Wavepower Ltd. The

---

* The project was also joined by Eire in 1979.

mid-1978 reference design is illustrated in outline in Fig 3.5.

A raft converter consists of a string of relatively shallow pontoons connected by hinges and moored in line with the prevailing wave direction. Power is extracted from the relative angular movements of the adjoining pontoons with the passage of waves underneath them. The first model tests in wave tanks were conducted with up to seven pontoons in the overall raft. It is likely, however, that the optimum configuration will be a string of three pontoons: the first two of equal length and the third (rear) pontoon twice the length. Such a configuration gives approximately equal sharing of power between the two lines of hinges which join the pontoons together.

Experimental work has shown that optimising against energy extraction efficiency can lead to a design which is capable of extracting energy from waves of periods from about 4–15 seconds with efficiency greater than 60 per cent. Thus the system can be made to operate effectively as an oscillator over a wide band—a desirable feature—the fact that the efficiency falls rapidly for the very large waves with periods over 15 seconds can be regarded as an advantage from the power take-off point of view. At present a complete full size raft (annual mean power rating, 1 MW of electricity) is envisaged as having dimensions 100 m × 50 m (the latter dimension is the one presented to the oncoming wave fronts) and 7–9 m deep. The size has yet to be optimised against overall costs in the light of more detailed information: smaller rafts would have a lower power output but this could well be offset by lower manufacturing costs of the main components.

For the primary power take-off the relative angular motions of the pontoons about the hinges would be used to pressurise water in a hydraulic main which would lead to a turbine driving a generator. Several different mechanical systems for producing the pressurised water have been studied and sea water is preferred as the working fluid. In the 1978 reference design a system of toothed racks and gears drives a set of ram pumps. The angular motion of the hinges varies with the sea conditions, but the stroke of the pumps is fixed by the throw of the driving crank. Thus the gear driven pumps can be of efficient design, executing one stroke for a small hinge rotation and several strokes for a large hinge rotation. Other pump systems are under investigation as offering much simpler layouts. Some further general discussion of the power take-off is given in Chapter 6.

The choice of constructional material would be in principle between reinforced concrete and steel. A complete raft of the size noted above in which most of the structure consisted of reinforced concrete would contain about 18,000 tonnes of the latter. Optimisation of the design could lead to a mixture, with the front and rear pontoons being of concrete and the middle one (which houses the power plant and is therefore more complex) of steel. The moorings of both an individual raft and of adjacent rafts in relatively close proximity presents significant problems to which attention is being given.

The concept of the converter is particularly susceptible to the problems of slamming: wave slam is the impact phenomenon which might occur when the leading pontoon re-enters the water if it were thrown clear of the sea by an especially severe wave. The loading incurred on the structure would be localised and short-lived, but could be very high. However, the problem can be avoided by careful design.

Progress and outstanding problems. Considerable progress has been made in understanding the theory and practice of the Raft concept, in which the work of Wavepower Ltd has been complemented by analytical work at the CEGB's Marchwood Laboratory. Model test results have been applied to mechanical and structural design: hulls have been designed in both steel and concrete and outline consideration has been given to factory assembly lines for quantity production. Several power take-off systems have been evaluated. The general design work has been supplemented during 1978 by open water trials of 1/10th scale models in the Solent. The main objectives of these were to gain sea experience and to obtain data, especially on mooring, in real sea conditions. A high degree of instrumentation was incorporated in the models.

From a strategic viewpoint, the Raft has a substantial advantage over the Rectifier and the Oscillating Water Column discussed above in that construction of the pontoons could take place at a wide range of sites and towing them on to location would be straightforward. These factors must help to reduce the costs. However, there remains a considerable design complexity in extracting power by means of large forces passing through oscillating mechanical linkages, which may prove difficult to overcome with cost-effective solutions. Further effort in the immediate future will be needed on two especially important cost centres in the design: the hinges and associated power take-off, and the problems of mooring.

Duck
The Duck, invented by S H Salter and currently being studied by Sea Energy Associates and at Edinburgh University, is generally regarded as an elegant concept, compact and economical in the use of constructional materials, but relatively complex. The complexity on the one hand can raise problems of system behaviour and reliability, but on the other hand can provide opportunities for a range of potential solutions to the design problems.

The essential features of the Duck are a very long floating cylindrical spine and a series of cams, or ducks, which are located on and rotate around the spine, see Fig 3.6. Power is generated from the motion of the ducks relative to the spine. The profile of the front face of a duck is chosen so that for relatively small rotations the displacements match those of the water particles in a wave. The rear face of a duck is circular and therefore does not displace any water. Theoretically, if the restraining force applied to the ducks is chosen correctly, an incident wave is unable to distinguish between a duck and adjacent water, and can transfer all of its energy to the duck: in this case the duck can be a perfect absorber.

The original concept of the spine was that of a continuous backbone which would obtain its stability from the self-cancelling of wave effects over several wave crest lengths. It provides also an advantage of allowing the power from many individual ducks to be collected together in a relatively straightforward manner up to a convenient level for generating electricity, thereby avoiding the problems of close mooring associated with, say, a multiple array of smaller rafts.

However, the analytical work which has been completed shows that the spine presents one of the most difficult design problem areas. Long spines, designed to remain rigid, would require impractically large diameters in order to accommodate the operational stresses. Several possible approaches to the problem are being considered, including articulated and curved spines. The articulation involves the design of discrete joints or hinges which, for maximum efficiency, will remain stiff in moderate seas but flex in heavy seas.

The main design options for the primary power take-off are discussed in Chapter 6. Whilst a number of systems has been investigated, many problems still remain to be tackled. In particular, there are practical difficulties associated with reversing high pressure hydraulic pumps and large strains in the high pressure mains, and achieving a layout of all components which will provide the right level of accessibility for maintenance will require a high degree of engineering ingenuity.

Careful thought has been given to the details of the shape of the ducks, and recent work has indicated that a profile of the top surface is possible which can help to ensure the survival of the system under the impact of very large waves.

Progress and outstanding problems. Considerable knowledge has been gained on the principles and behaviour of a number of duck and spine designs by several routes:

—the construction of a new wave tank at Edinburgh which can model any required wave conditions;
—experimental work in the wave tank on a model duck subjected to all the weather conditions of the North Atlantic, including breaking waves. The single duck was mounted in a rig which simulated the behaviour of adjacent ducks, and the development of this rig was an important step forward in providing the ability to simulate a string of ducks under controlled conditions in the laboratory;
—testing models in the tank under conditions of very steep waves, which indicated that slamming behaviour is likely to be acceptable;
—the testing of a 1/15th scale model duck in the National Maritime Institute (NMI) wave tank at Feltham, which showed good correlation with the 1/100th scale work at Edinburgh, thus giving confidence in the validity of further work at the small scale;
—the construction and testing of a 1/10th scale model of one design of duck and spine in Loch Ness.

However, engineering solutions which are accepted as credible are not yet available to the design problems of the spine and the power take-off. They require much further effort with special emphasis on:

—survival under the worst conditions;
—the problems of maintenance;
—engineering cost optimisation.

3.2 Alternative concepts

Over many decades the possibility of tapping the power of ocean waves seems to have exerted a peculiar fascination over inventors. Hundreds of ideas have been put forward. After a detailed analysis by the National Engineering Laboratory, four concepts were adopted for initial study under the Department of Energy programme. However, a steady flow of new proposals is received by the Energy Technology Support Unit which acts as the primary point of contact with the proposers: over 40 were considered during 1977.

The Wave Energy Steering Committee set up a Technical Advisory Group (TAG 1, see Appendix 3) to assess and advise upon new concepts, and to initiate work on those which show promise or can fill a gap in the wide spectrum of technological types which should be studied. It is by no means certain that the most acceptable concept lies amongst those which are in the programme at this stage. The Committee is keeping an open mind until the analytic and experimental work enable a clear choice to be made.

The important criteria in assessing the merits of new ideas are the likely engineering feasibility and cost-effectiveness. However, the appropriate data on costs are usually at best very approximate and often absent. Other factors are therefore taken into account as clues to the main criteria:

—hydrodynamic efficiency in 'mixed' seas;
—feasibility of the proposed power take-off (including mechanical efficiency, component life under the arduous conditions, access for maintenance, etc);

—provision of a suitable frame of reference against which moving parts can react, or a valid system of force cancellation;
—survival in extreme wave conditions;
—compactness (which has a direct bearing on both cost-effectiveness and environmental impact);
—mechanical and structural simplicity.

Most concepts which have been assessed failed on several of these factors, but two have been recommended for initial financial support from the Department of Energy.

One possible very simple way of classifying the wide variety of potential concepts, so that efforts can be made to ensure that no major avenue remains unexplored, is based upon three essentially geometrical properties:

—width (wide or narrow in relation to the average wave crest length);
—length (long or short with respect to wave length);
—non-directional or directional (able, or not, to absorb energy from several or all directions simultaneously and with approximately equal efficiency).

There has, for instance, been considerable theoretical interest in the idea of a 'point absorber'—a converter which would be narrow, short and non-directional—as a possible important gap in the original programme.

Long converters would be easier to moor than short, wide converters and ease the problems of power collection (important considerations in the overall cost) but these features can be counterbalanced by short converters tending to be more effective in capturing energy on the basis of unit length of working face. The final choice of the best system overall will involve engineering judgement, compromise and optimisation over a considerable range of parameters, some of which will be in conflict.

A 2 × 2 × 2 classification of concepts based on the three properties noted above is shown in Fig 3.7. Two of the eight categories are void, since a wide or long converter cannot be non-directional. It is of interest that all four designs chosen for the mainstream of the original programme fall into one particular category. The two new concepts on which the exploratory work is now included in the programme, described below, are in alternative categories.

Non-directional

	Narrow	Wide
Short	The Duct	✕
Long	✕	(Ring absorbers)

Directional

	Narrow	Wide
Short	Single Raft Single Rectifier Single OWC Single Duck	Multiple Rafts Multiple Rectifiers Multiple OWC Multiple Ducks
Long	Tube-pump Japanese Kaimei	(2-dimensional array of point absorbers)

Fig 3.7 *Classification of converters by three geometrical properties*

Vickers Duct

A funding contribution towards a proof of concept study has been allocated to Vickers Ltd who put forward a design concept which is based essentially on a resonant oscillating water column but which is novel in the nature of the damping applied to the column and in the fact that it may have some of the characteristics of a point absorber. The concept is illustrated in Fig 3.8.

The philosophy which led to the creation of the concept included the following important points:

—survival in heavy seas is more easily assured if the converter is submerged (it sits on or near the seabed);
—economy and reliability will be more readily attained by adopting the simplest possible power take-off machinery (a water turbine working in unidirectional flow);
—apart from the turbine/generator, a minimum of moving parts.

The mode of operation can be described in relation to Fig 3.9. A submerged tube is attached to a closed air reservoir at one end with the other end open to the sea.

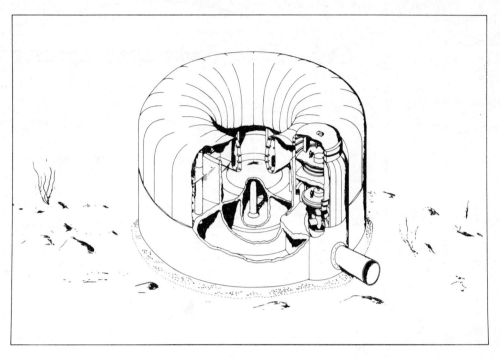

Fig 3.8 *The Duct (a new concept by Vickers Ltd)*

The water in the tube can oscillate in response to changes in water pressure caused by surface waves passing over it. The water column's length is chosen to relate its natural period to that of the waves, so that resonance occurs and the oscillation is amplified; the restoring force is provided by the pressure in the closed air reservoir.

When the water column oscillates, the amplitude is sufficient for water to overspill into the reservoir for a number of cycles. This raises the reservoir water level, increasing the pressure in the reservoir and thus depressing the mean level of the water column's oscillation; overspilling stops, although the water continues to oscillate. If an outlet valve on the reservoir is opened, water is discharged, the pressure is reduced, the mean level of oscillation rises and overspilling starts again. By balancing inflow and outflow, a steady state can be maintained, with water being discharged continuously. Power, represented by the product of the water discharge rate and the pressure differential, can be extracted by placing a turbine within the discharge flow.

The scale of the converter will be determined by the requirements that the air store (basically acting as a spring) and the water collecting trough will need to be large. In addition, to attain a ratio of scale to power output comparable to other concepts the water column will need to capture power over several times its own width. These problems may, on further investigation, prove to be insuperable, but in the meantime the basic simplicity and ruggedness of the concept are attractive.

Flexible bag

This concept was invented by Professor French, Lancaster University, and the early work on it was funded by the Science Research Council. The Department of Energy has recently accepted responsibility for further funding.

The design, illustrated in principle in Fig 3.10, comprises a number of air-filled bags attached along the top of a submerged hull lying head-on to the sea. The bags are formed by dividing a long tube made from flexible material about 8 mm thick, by parti-

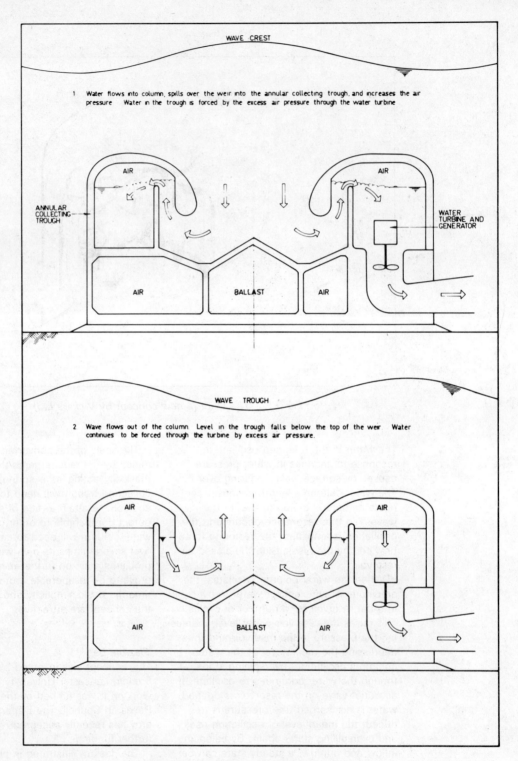

Fig 3.9 *Vickers device working cycle*

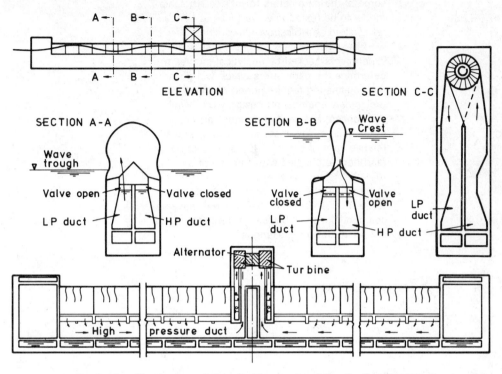

Fig 3.10   *Illustration of the working cycle of the flexible bag*

tion membranes of the same flexible material.

The hull is a narrow prestressed concrete structure, typically about 190 m overall length, 6 m beam, and 8 m deep for an output of several MW, containing high and low pressure air ducts connected to the bags by non-return valves.

The ducts are connected together through a pair of air turbines located in a machine housing midships. Together the bags, ducts and turbine comprise a closed circuit air system. The main part of the hull lies submerged with its base some 13 m below mean wave level. Raised tank structures at each end of the hull provide trim and reserve buoyancy.

In operation the converters would be moored in line with the direction of the incident waves and side by side, spaced sufficiently apart to swing clear of one another under the most adverse weather conditions.

As in the Japanese Kaimei vessel described earlier under the Oscillating Water Column, the vertical planes converting the wave energy lie normal to the wave motion, the waves being attenuated as they travel through channels formed by adjacent converters. As the wave trains run along a converter the rise and fall of the water on each side of the bags causes the latter to act as bellows pumping air from the low to the high pressure ducts. The differential pressure between the high and low pressure sides is governed by the power take-off, and is of the order of 15 kN/m^2 above the datum pressure.

The concept is unusual in that it was not conceived as a resonant device. The surface from which the wave energy is extracted is a light flexible membrane. The motion of this membrane is such that the hydrodynamic added mass is also very small. This means that the bag does not have a resonant period within the wave spectrum. However, the added damping (the ability to radiate waves, or conversely absorb waves) is high and the device could in theory have a high efficiency over a very wide bandwidth, that is covering most of the wave spectrum. The converter will not therefore exhibit the same characteristics as the other floating converters.

The present form of the concept has been evolved following a range of static single- and seven-cell model tests up to 1/12th scale which allowed optimisation of some of the parameters for a 1/40th scale model to be tested in wave tanks at Lancaster, Salford and Glasgow Universities. The work is still in its early stages: small scale experiments backed by analytical studies to determine the parameters which will govern the engineering requirements, coupled with preliminary engineering design work. Whilst the concept has several inherent attractions, such as low materials content and relative ease of construction and mooring (and hence, coupled with the use of air turbines, relative cost-effectiveness), ensuring the integrity of the air bags in relation to the overall capability to survive in adverse sea conditions will be one of the most important design challenges.

4 General aspects of structural design

The successful design of a wave energy converter will involve the prediction of water-induced motions and loads and structural response, based on experimental data and theoretical analysis. Structural design will require application of established design procedures and data for steel and concrete structures, contained to some extent in existing codes for ships and offshore structures but probably requiring adaptation for wave-energy converters. Problems influencing the choice and deployment of materials, in particular problems of corrosion and fouling, will also require careful consideration. This chapter is concerned with the development of analysis methods and procurement of experimental data relating to fluid loading and structural response of devices, together with assessment of some major materials problems and structural design procedures. These problem areas have been the responsibility of Technical Advisory Group No. 3 of the Wave Energy Steering Committee (see Appendix 3), on which the device development teams are represented.

The design of any marine structure exposed to wave action requires the prediction of:

—extreme wave-induced loads;
—histograms of cyclic loads which may cause fatigue damage.

Ideally both these items of information should be defined statistically, corresponding to a statistical description of the wave conditions (significant heights, mean periods and directional spreads) which the structure will experience during its life. Impulsive forces caused by breaking waves and by slamming of the structure may also contribute substantially to extreme wave-induced loads.

For structural analysis purposes loads can be classified usefully into three major categories:

—primary bending and twisting moments and direct and shear forces acting on a major structural component (for instance a Salter spine or a Cockerell pontoon), causing deformations and stresses throughout the component;
—secondary loads, for instance local hydrodynamic or inertial forces, which contribute to the primary loads but also cause localised deformations and stresses;
—forces and moments transmitted at mooring points or at connections between the elements of a multi-component array.

In each case the wave-induced loads will be superimposed on still-water loads which are caused by an imbalance in the distribution of the weight and buoyancy forces. There is a very extensive literature on the topic of wave loads on marine structures, including both ships and off-shore structures, but unfortunately many wave energy converter designs have features which present special problems not accommodated by the established methods. A key difference is that in many cases they involve a number of major moving components and must be treated as multi-element bodies in the analysis, responding to wave forces with more than six degrees of freedom.

4.1 Evaluation of wave-induced motions and loads

For analysis purposes the response of a floating structure to wave energy action is usually assumed to be linear, i.e. coefficients in the equations of motion are constant and motions, loads, deformations and stresses are proportional to wave height. In some cases hydrodynamic coefficients may be obtained from established theoretical or empirical data for simple cylindrical, spherical, ellipsoidal or box-shaped bodies. For more general geometries, hydrodynamic pressures on elements of a submerged surface may be computed using diffraction theory: integration of pressures over the complete submerged surface then yields hydrodynamic forces acting on the floating body as a whole.

A large number of computer programs and analytical solutions is available. Such programs, especially those based on three-dimensional theory, are unfortunately very expensive to run and, although they will have an important role to play in the further development of wave energy converters, for many practical purposes simpler approximate methods are needed.

It is likely that such linear analysis methods will be of particular value in the following problem areas:

—evaluation of the performance of alternative converter configurations in moderate wave conditions;
—evaluation of cyclic load histograms in relation to fatigue failure;
—as a starting point and a basis for semi-empirical prediction of motions and loads in designing against extreme operating conditions.

However there are the following limitations:

—the response of most of the converters, particularly in extreme operating conditions, will be at least partly dependent upon non-linear effects (in the relationship of the coefficients in the equations of motion to wave height);
—no account is taken of the effects of impulse forces caused by breaking waves and by slamming;
—the statistics and spectral form of extreme wave conditions are not known reliably.

Whilst the theoretical methods, including possibly those of non-linear analysis based on hybrid computers, will be valuable in the further development work, the evaluation of response under extreme conditions is likely to have to rely heavily upon experimental data from small-scale tank tests and larger scale sea trials. Becuse many of the problems involved are similar to those relating to mooring the description of the latter in Chapter 5 arrives at a similar conclusion.

Under the guidance of the Technical Advisory Group, progress has been made on a number of problems in this area during the two year feasibility study in a number of laboratories. Work carried out by the British Ship Research Association (BSRA) and the National Maritime Institute (NMI) on the theoretical analysis of wave-induced motions and loads has included:

—an assessment of existing linear analysis methods and computer programs for ships and off-shore structures;
—an extension of the existing methods and programs to deal with two- and three-dimensional representation of wave energy devices;
—a comparison of the computed responses with experimental data for the wave energy converter modelling work.

Comparison of the theoretical and experimental results has been encouraging and, while the limitations of linear analysis are recognised, it is believed that the BSRA and NMI programs now provide a valuable means of assessing the performance of converters and of developing the designs further.

Significant theoretical studies in this area, with particular application to ducks and rafts, have also been carried out by the CEGB. Information on the wave induced loads in the spines of the duck design, pointing to a need for the articulation of the spine as a means of avoiding large bending moments, has been obtained from a theoretical study carried out under contract by the National Engineering Laboratory.

Some limited progress has been made in experimental evaluation of wave-induced loads and stresses in ducks and spines, and at hinges and mooring points for rafts. Experimental trials in Loch Ness and the Solent were subject to delays by unfavourable wave conditions. However, realisation of the continuing usefulness of small scale work has led to the construction of an advanced wave tank at Edinburgh University in which models can be subjected under realistic conditions to all forms of wave regime appropriate to the North Atlantic. As a result of the successful commissioning of this tank, it is likely that much further small scale work can be done which is not subject to the lack of control over the weather and the difficulties of making measurements in real open water.

Theoretical studies of certain aspects of the impulsive loads caused by breaking waves and slamming effects have been carried out at Cambridge University, the National Maritime Institute and the Hydraulics Research Station. Comparison with the data for ships (Lloyds Register) and offshore structures has demonstrated the potential importance of these impulsive loads. Severe breaking wave conditions have been produced in controlled model experiments in the wave tank at Edinburgh University.

Many of these theoretical approaches to the analysis of the wave-induced motions and loads are capable of considerable further development, which will be needed to enable them to be applied with greater confidence and precision in the continuing design work. The Technical Advisory Group is planning a programme of these further studies. In addition the experimental evaluation of wave induced motions and loads and of the structural response should proceed further along two main routes:

— systematic experiments on small scale models of each design in wave tanks under controlled conditions, including regular and irregular long-crested and short-crested waves and including extreme and breaking waves; this is a priority item which will require additional wave tank facilities;

— further tests of the Loch Ness and Solent type at intermediate scale to provide a check on the scaling effects and experience of 'going to sea' prior to subsequent larger or full-scale trials: the modelling of structural response is likely to be more effective at intermediate than at small scale.

4.2 Evaluation of structural response

For the purpose of evaluating structural response (that is, deformations and stresses) fixed or floating converters may be divided into two basic categories:

— 'flexible' systems, such as a set of ducks mounted on a long spine, in which the structural deformations are likely to be coupled strongly with the overall wave-induced motions;

— 'rigid' systems (including possibly a chain of rigidly constructed raft pontoons) in which the structural deformations are effectively uncoupled from rigid body movements: in such cases structural response can be analysed separately from rigid body motions, using loads derived from the latter analysis.

A wide variety of methods, varying from simple formulae to complex finite element analysis embodied in computer programs such as NASTRAN, SESAM and ASAS, is available for performing this type of analysis. However, reliable and economic evaluation of structural response does require considerable judgement in the choice and application of the particular analysis methods. An evaluation has been commissioned at Lloyds Register of the dynamic structural response of wave energy converters using NASTRAN.

Other studies on structural design which have been initiated include:

— preliminary work by Lloyds Register in collaboration with the engineering consultants and the device teams, of a provisional set of Guidance Notes on structural design which might serve as a basis for an eventual Code of Practice and which would in the meantime provide practical guidance on structural design;

— work by British Shipbuilders to develop and prepare cost estimates for reference designs in welded steel of the raft and the oscillating water column for comparison with alternative reference designs by Rendel, Palmer & Tritton based upon construction in concrete;

— Rendel, Palmer & Tritton have been commissioned to evaluate the possible structural applications of glass reinforced plastics in wave energy devices.

Further work in the near future will also need to cover the identification and detailed analysis of critical structural components. In due course for those designs which are firmed up the proof testing of critical components, for instance hinges in the rafts and the joints in possible articulated duck spines, under static cyclic loads, will need to be carried out to ensure the long term durability at sea: this is likely to be an expensive item in the forward programme.

4.3 Structural materials

One of the most significant decisions to be made in the design is the choice of the main structural material—concrete or steel—a problem which parallels the situation in the offshore oil and gas rig industry, where production platforms have in fact been built in both materials. The operating conditions are sufficiently dissimilar between the requirements of wave energy and offshore structures, however, that by no means all of the information from the offshore research programmes and the resulting operating experience will be directly relevant in the wave energy field.

Sea water is a very corrosive medium, the most serious effects probably occurring in the so-called splash zone where the surfaces of materials are exposed intermittently to water and air. Corrosion effects will be an important consideration in the design of wave energy converters, especially since one of the criteria must be to achieve long operating life with the minimum of maintenance.

Concrete

Some valuable experience already exists which illustrates that concrete structures can survive for long periods in the marine environment. For instance the Tongue Sands Fort, built of reinforced concrete with an upper steel deck some 13 km off Margate, is yielding information on the effects which can occur over an exposure of some 35 years.

Considered as a structural material, concrete has the advantages of being readily formed in complicated sections and of maintaining consistent quality in the process. Its negligible tensile strength is overcome by the use of reinforcement or by prestressing, whilst its compressive strength (although not of the same order as that of steel) is high enough to result in very economical compression sections. Over recent years, engineers have sought continuously to employ concrete in new situations and to increase its durability and economy in more familiar applications. It is being exploited in a range of structures (dams, bridges, nuclear containment vessels, tall chimneys, concrete ships, offshore gravity platforms, etc) which subject it collectively to most of the conditions which will apply in wave energy converters. One notable new condition, however, will be the fatigue loading due to the wave action.

Much valuable design information will become available through the Department of Energy's programme 'Concrete in the Oceans' which was started in 1976 in collaboration with about 20 companies. This is aimed at providing knowledge to improve the design, construction and long-term performance of concrete oil production platforms. Many aspects of the behaviour of 'standard' reinforced concrete will be determined adequately in this programme.

However, concrete in modern offshore structures has proved to be an expensive material, and wave energy converters can almost certainly not sustain the costs that are incurred in meeting oil company specifications. Nevertheless, there will be important differences in the manufacturing conditions: wave energy converters would be produced in large numbers using production line techniques leading to lower costs. Success in the wave energy programme will depend on achieving a relatively cheap concrete structure which still retains adequate mechanical properties and corrosion resistance. The 'Concrete in the Oceans' project is currently under review and may well be extended to cover additional requirements from the wave energy programme.

A specific aspect of wave energy converters will be the associated large electrical currents and voltages arising from the electricity generation and transmission processes, which may give rise to stray current problems in reinforced concrete. A preliminary study of possible effects and solutions is under way.

In summary, concrete is a very good structural material and most of the further work necessary to advance the state of the art in the direction of lower costs whilst maintaining durability is likely to take place outside the wave energy programme. The situation is kept under review for the Device Teams and the Wave Energy Steering Committee by the consulting engineers and materials specialists.

Steel

Steel is a possible alternative to concrete as the main structural material but even if it were not chosen for this purpose many individual components would incorporate it. Steel structures have been used successfully for many years in marine conditions, and there is a vast fund of knowledge and experience of using corrosion-resistant coatings and/or cathodic protection. Some commercial shipping experience does not look promising, but poor performance of anti-corrosion systems can often be attributed to a choice of inferior materials, insufficient care over the conditions of application, or misuse. Ships with good coating systems can now anticipate a life of some four years between dry dockings and much longer between major repairs to anti-corrosion coatings. Since the wave energy converters may have to be taken off station for other maintenance or refit reasons over such intervals, it would appear that the corrosion aspects need not be the limiting factor.

Mechanical damage to coatings will occur due, for instance, to impact with flotsam or with service boats. The wave energy programme should be able to benefit from the development of new paints for application to wet surfaces, which is being stimulated by the need of offshore platforms.

An area of some ignorance, should steel be chosen, is fatigue failure. The wave loading on structures, predicted as in the earlier part of this chapter, both in magnitude and frequency, can be severe in terms of the possibility of inducing failure due to cyclic stresses. Since complex welding joints are of crucial importance in steel offshore oil/gas platforms, the subject is receiving increasing attention in that field. The Department of Energy supports a major research programme to provide quantitative data for the assessment of the safety and reliability of steel structures in the marine environment (the UK Offshore Steels Research Project). Whilst, unlike the situation on concrete, much of the work is designed for direct data gathering under conditions specific to offshore platforms and not relevant to the geometry, loading and other conditions of wave energy converters, some results of general significance are beginning to emerge, supplemented by other programmes such as Admiralty work on the fatigue behaviour of gas turbine alloys and work at the British Non-Ferrous Metals Research Centre on other non-ferrous alloys.

The progress in all these programmes is being monitored for the relevance of their results to wave energy devices.

Other materials

Early work towards reference designs for the converters has contemplated the use of rubbers, in some cases as vital components in addition to their use as sealing materials against the ingress of water.

Whilst, in general, the use of various forms of vulcanised rubber as sealant will present no major new problems, some designs may need to incorporate large circumferential seals which are beyond present experience and would require extensive testing work.

Rubber may be incorporated as an essential structural material—for instance as hinges for the flap gates of the Rectifier, or in the form of tyres as the power take-off for some designs of the Duck. However, from the materials point of view it would be necessary to extend the present range of knowledge of the behaviour over long periods under seawater, in particular there is a lack of information for the prediction of failure rates from fatigue under cyclic stressing. The strain analysis of large rubber components may not be calculable to a sufficient degree of certainty and full-scale trials would be needed. These are being considered in the current extension to the programme.

There has been some interest in the possible use of glass-reinforced plastics as structural material, based upon the growing experience of their successful application in small boats. One of the main attractions would be their very good resistance to corrosion, but there may be disadvantages due to a loss of strength over long time periods due to attack by moisture and to dynamic fatigue. For existing applications this is not critical since low working stresses can be employed, producing structures which are durable and resistant to creep. To render the use of such materials cost-effective in the wave energy programme would probably require better design techniques and more adequate data.

Marine fouling

Fouling is the settlement and growth of marine plants and animals on any part of a marine structure. Other floating debris which comes into contact with a wave energy converter could also be included as fouling. Extensive fouling of the wave energy converters may well be unavoidable, and the functioning and durability of the mechanical components could be affected seriously without proper attention to design details. The possible effects of marine fouling are shown in Table 4.1.

Table 4.1 Possible effects of marine fouling

Increased weight of structure
Increased volume of structure
Increased surface roughness and drag
Masking of surface to obviate routine inspection and maintenance
Removal of fouling often removes protective layers as well
Changes in corrosion fatigue behaviour
Enhanced probability of brittle fracture
Tribology effects on moving parts, eg reduction in gear and bearing lifetimes
Blockage of pipes, valves and gates
Prevention of proper sealing of moving seals and valves

The phenomenon has been investigated for many years, mainly in connection with the effects on the fuel consumption of ships due to the increased drag and friction, but also mooring ropes and chains and navigation buoys have been examined. More recently there has been a much increased interest in the effects of fouling on platforms in the North Sea, where large diameter increases have occurred on cylindrical components. A Department of Energy Working Party has been formed recently to advise on the need for action with respect to platforms. The wave energy programme will be able to benefit from the experiences and work reported to the Working Party.

A particular item in Table 4.1 which is of concern to the offshore platform operators is the possible influence of fouling on corrosion and corrosion fatigue, and this is receiving attention.

Few data exist on the types of fouling to be expected in the possible locations for wave energy devices off the Hebrides, and an experimental programme has been initiated. Experience of the Harwell Laboratory and the Scottish Marine Biological Association has been combined to design and build an experimental test rig positioned about 10 miles west of South Uist. The experiment will monitor fouling on standard test panels from the first settlement period (spring) through at least two summer growth seasons at three depths and at the seabed.

That serious problems may be encountered is illustrated from the fact that general data on fouling by barnacles and mussels indicate a possible buildup of some 40–50 $kg/m^2/year$. The increased downthrust on fullscale wave energy devices could be in the range 1–2 tonnes/year/m of wave front. Whilst the significance of this will differ for individual designs, it is too large to be ignored. A choice of remedies may be possible in principle.

Preliminary consideration is now being given to the possibility of large scale fouling tests in an area suitable for wave energy development, in the form of a large moored floating structure which may well be valuable for other test purposes such as mooring systems. The main reason for such a large scale test for fouling would be to determine whether the particular types of organism responsible for fouling would behave in the same way on a full scale wave energy device as the way they behave on small test panels. There may be some reasons why the behaviour could be different.

Finally it should be noted that siltation can be regarded as a form of fouling, and this is of particular relevance to the possibility of a converter such as the Rectifier mounted directly on the seabed. The importance of the transportation of sand and kelp by the waves into the Rectifier can only be determined after a detailed design is available and a specific location has been chosen. For many locations off the Outer Hebrides, fouling by kelp might be serious: see Chapter 8.1.

5 Mooring

The ultimate safety and ability to survive of a floating wave energy converter will depend upon the provision of a mooring which can withstand the most severe storms likely to be encountered in the sea area in which it is situated. It would not be tolerable that a significant proportion of the country's energy supply could be destroyed, or even disabled, in one extreme storm.

Until comparatively recent times, almost all the development and practice of mooring have been associated with ships in relatively sheltered waters. Exploration and production for oil and gas in offshore locations created new requirements for mooring in more exposed conditions. Whilst the design of wave energy converters will be able to lean on the experience and skills developed for the offshore industries, the conditions of operation will differ sufficiently to introduce additional requirements which are beyond the present state of the art if the desired degree of reliability is to be achieved at acceptable cost.

Preliminary work in 1978 by the engineering consultants to the Wave Energy Steering Committee indicated that mooring costs could become a substantial proportion of the total cost of a wave power station in some designs. Thus both reliability and costs are vitally important considerations in the design of the mooring system. The fact that a large number of moorings would be required ultimately should bring economies of scale both in initial cost and in maintenance using dedicated service vessels and equipment.

Whilst the detail of the design will depend upon the particular wave energy converter to be moored, it is possible to pursue many of the problems in a more general way. A Technical Advisory Group has been set up to provide expert knowledge to WESC and the engineering teams developing the converters, and to initiate an appropriate research and development programme (see Appendix 3).

The first approach taken by the Technical Advisory Group was to examine how the present state of mooring knowledge and component availability could be applied to the tentative reference designs of converters so far available. There is no doubt, however, that as development proceeds on both converters and moorings close integration will be necessary in order to arrive at the optimum solution. For instance, designing the converters for lower mooring forces may not *ipso facto* produce lower costs overall: higher conversion efficiencies along with higher forces might well be more economic. The current philosophy and the development of the initial general programme on mooring by the Technical Advisory Group are described below in Section 5.3 after a brief description of the mooring systems from which a choice may be made.

5.1 Mooring systems

In considering all the possible types of system by which wave energy converters could be positioned it is advisable to include not only conventional mooring but also forms of dynamic positioning which are in use or under development by the offshore oil industry. The main systems which have been reviewed are:

Dynamic positioning
- by propulsion
- by active mooring

Mooring
- by tethered buoyant platform
- by catenary
- by taut elastic systems

Dynamic positioning by propulsion

This technique consists of an array of carefully positioned propellers or thrusters, each controlled from a central computer system to maintain position over a predetermined fixed location on the seabed: an array of sonic transducers on the seabed is usually required as fixed reference points.

The present development appears to be limited to sea conditions which could be considered as the worst annual storm rather than the 50-year or longer term categories which are important for the survival of wave power converters.

In severe storm conditions, more power is likely to be needed for positioning than may be available from the converter itself, and costs would be very high. Therefore the further development of this system for wavepower generators has not been recommended.

Dynamic positioning by active mooring

This system is used for positioning and moving pipelaying barges during the laying operations. It comprises up to 12 wire hawsers, each with an anchor, spread in an array around the barge. The inboard end of each mooring line is held on a winch which is servo-controlled, usually by a computer.

The basic design requirements differ significantly from those of a wave energy converter. Whilst there would be some advantages for the latter in terms of reduced cable and anchor specifications, expensive machinery and control equipment would be necessary, requiring frequent maintenance. Moreover, substantial guaranteed power would be needed to ensure the safety of the wave energy converters in storm conditions.

Tethered buoyant platform mooring

This is another recent development—mooring lines under continual tension (and usually near vertical) react against the buoyancy of the moored structure. It is under serious consideration for a future generation of offshore oil/gas production platforms, although full-scale experience is not yet available. The system is being developed to eliminate the heave motions, normally associated with catenary moorings, to avoid damage to the oil production risers. In practice, for large structures, a relatively expensive arrangement involving a number of 'tethers' is likely to be required.

A substantial R and D programme has been initiated by several large oil companies and their contractors. At present the results are confidential to them, but it is expected that the technology would be made available for other uses when the system has been confirmed on an offshore oil field. However, since the moorings for wave energy converters will need very high compliance it is unlikely that this type of system will be applicable.

Catenary mooring

Undoubtedly this is the system on which most experience exists, and which is being used increasingly in more exposed conditions for relatively large structures (larger than wave energy converters) though rarely designed for extreme conditions. It provides the near horizontal pull necessary to ensure good performance from 'drag' anchors and provides 'spring' in the system. Both characteristics tend to require a mooring line with high weight to unit length.

Catenary moorings are currently being used on the semi-submersible platforms on the Argyll field, and the Buchan field is about to be developed with a similar system.

Design requirements will differ in the case of wave energy converters—for example, station keeping may be less critical— but ultimate survival may be required in more severe conditions and over longer periods without maintenance and inspection.

Taut elastic mooring

This type of mooring relies on the very high elasticity of man-made fibre ropes, rather than the catenary shape of a heavy line, to provide the necessary 'spring'. The ropes are neutrally buoyant, or nearly so, and would be essentially straight. This type of mooring has not previously been adopted in its pure form for exposed moorings, but is widely used in berthing situations.

Choice of system

Whilst the catenary system was chosen initially for more detailed consideration on the assumption that simplicity of concept together with known experience would be most likely to lead to an acceptable system, attention is being given also to other systems which are under development for different applications. For rigidly held converters, the cyclic forces at wave frequency would be large and would totally dominate all other forces. The forces in the catenary could be reduced to an acceptable level by allowing the converter to move in each wave: a compliant mooring would have many advantages, combining the desired degree of compliance with compactness (the latter feature having the important effect of allowing the converters to be moored relatively close together). Such moorings will need development. In the event, a combination may be needed for an array of converters, depending in detail upon the converter design and the location.

5.2 Mooring components

There are three main components to a mooring system:

- —anchors;
- —mooring lines;
- —shackles or joining components.

Anchors

There is currently a range of drag, deadweight and piled equipment and techniques available for anchoring, the largest sizes and the forces they have to resist being determined by the present market requirement. In addition experiments have been carried out on various embedment anchors, but presently at small scale.

The dominating parameters which govern the performance of any anchor lie undoubtedly in the soil composition of the seabed. If rock is at or near the surface then the choice becomes very limited to either rock piles or deadweight anchors. The efficiency of all other current means of anchoring is totally dependent on soil conditions and before any assessment of anchor holding capacity, hence size, can be made, accurate and extensive core sampling or tests must be made for each particular site. It is therefore the case that there is no single anchor which will be suitable for all locations. In considering a large wave power installation made up of many modules extending for possibly hundreds of kilometres, it is likely that a large 'wardrobe' of anchors will be necessary to be cost effective and to meet the load requirements on each particular anchor site.

The largest size of individual anchor which currently exists is at present considerably smaller than that which it has been assumed would be required. The alternative of using several smaller anchors in place of a single unit may be possible, but may have economic penalties. The largest fluke anchors are around 70 tonnes weight with a predicted holding capacity in the region of 6.0 MN. At present there is no known reason to expect that adverse effects of increasing size and scale of fluke anchors exist, however this is a totally unexplored area and it may be found that increased scale would introduce problems in handling outside current experience. Most manufacturers contacted believe that there are no particular problems in manufacturing anchors of the sizes and scale envisaged.

Difficulties of laying and maintaining anchors during restricted weather windows off exposed coasts are liable to incur a heavy cost burden and it may well be that special self-burying anchor techniques (i.e. without surface support) will have to be developed if the ultimate potential of wave energy is to be exploited.

Mooring lines

There are three categories of mooring line currently available although strictly speaking only two are widely used for mooring to anchors. These are chain and wire rope.

The third, man-made fibre rope, is used extensively for above-water moorings of ships to quays and to single buoy moorings but few applications to the seabed exist. The properties that man-made fibre ropes have to offer make them worthy of serious consideration. Since the introduction of nylon fibre in 1940 and later polyester, polypropylene and polyethylene to the rope industry, the use of these fibres has grown dramatically to the extent that they are now used almost exclusively in place of the natural fibres coir, hemp, manilla and sisal. They are not susceptible to bio-degradation and the quality of the raw material can be carefully controlled. The increased strength offered by man-made fibres makes possible smaller more handleable sizes and their capacity for storing strain energy is high.

Chain is the traditional mooring line to anchors and is the most widely used throughout the marine industry. However, the offshore industry is tending to use wire ropes in increasing quantities to replace chain: the major advantage being ease of handling and weight saving, the equivalent strength chain is five times heavier. Chain is relatively expensive and although it can have a life of 20 years or more the life can be reduced to only a few months under adverse conditions. Its past development has been constrained by handling and storage considerations. With long maintenance-free life of increased importance there is scope for a new approach to design.

Table 5.1 gives a simplistic appraisal of the relative merits of each type of mooring line for comparison. No line is ideal in all respects, and the final choice will depend upon the accumulation of more detailed knowledge—especially of the load extension characteristics and the fatigue properties. An initial programme of fatigue testing of man-made fibre ropes under wet conditions is under way at the National Engineering Laboratory. Other problems presently being assessed include: corrosion fatigue of steel wire ropes, the validity of accelerated testing, failure at terminations and compliant moorings with taut elastic properties.

Shackles
Whilst there may be some difficulty in ensuring survival of shackles and swivels over long periods, no new basic problems are presented by wave energy converters. Shackles have been developed already up to the working load specifications likely to be required.

5.3 An approach to the mooring problems

The total force on the moorings will have components due to the waves, currents and winds. For the most likely locations, the wave-induced forces will probably be much larger than those from currents and winds, although in the final designs the latter cannot be neglected.

In evolving the design and specification of a mooring for an array of wave energy converters the following points must necessarily be considered:

—the design of the converter;
—the environmental forces expected at a particular location (in particular the maximum predicted wave height and tidal range);
—the depth of water and the composition of the seabed;
—the degree of excursion of the converter, possible or desirable:
 ● minimising the mooring forces requires consideration of the coupling between the motion of the converter and the mooring system;
 ● introducing a considerable degree of compliance.
In practice there will be restraints on the amount of excursion possible, principally from the electric cable connections but also due to the practical limits of compliance of which a mooring system is capable;
—the proximity of other converters;
—the likely incidence of marine fouling;
 ● this can vary considerably with location and can affect the forces and the performance of the mooring system itself (see also Chapter 4);
—the logistics of deployment and proving the system;
—the degree of difficulty and frequency of inspecting, testing and maintaining the system:
 ● a complete guarantee of integrity cannot be possible. There must be a balance between appropriate safety factors and the economics of the system. Prototype and early installations are likely to be over-designed until experience has been gained.

Until the final designs of large converters have been developed, some of these factors cannot be taken fully into account. However, to meet the present requirements of the wave energy programme several different but related approaches are being pursued.

Immediate programme
Three designs of floating converter either have been or are planned to be tested at approximately 1/10th scale in a sea environment. Conventional catenary mooring technology has been or is likely to be used for these tests. High safety factors or redundancy are incorporated to ensure that the

Table 5.1 Properties of mooring for wave energy converters

Material	Corrosion resistance	Fatigue life	Abrasion resistance	Susceptibility to marine fouling	Handleability	Termination	Compliance	Special features	Usage in similar applications
Chain	Poor	Unknown	Poor	Average	Difficult	Various	Catenary limited	—	Widely used
Wire/rope	Poor/good	Unknown	Good	Average	Good	Socket or eye splice	Catenary limited	—	Widely used
Nylon	Excellent	Unknown	Very poor	High	Relatively light	Eye splice	Very elastic	—	None
Polyester	Excellent	Unknown	Very poor	High	Relatively light	Eye splice	Very elastic	—	None
Polypropylene	Excellent	Unknown	Very poor	High	Relatively light	Eye splice	Moderately elastic	Buoyant	None
Parafil	Excellent	Good on limited data	Poor	Low	Relatively light and good	Patented metal termination	Slightly elastic	Buoyant	Very little

experiments do not fail at this information gathering stage.

To provide a sound basis for prediction of full scale mooring requirements, it will be necessary to carry out accurate simulations during tank tests and to confirm these with more accurate modelling at the 1/10th scale stage. The information so obtained will be important in correlating analytical predictions.

Mathematical modelling

Mathematical modelling for the design and assessment of mooring systems is seen as a necessary part of the overall programme, which must be dovetailed into tank and sea tests of models. Much of the recent work in this field has its origins in the offshore oil industry's requirements, and it is anticipated that the work needed for wave energy purposes will be integrated wherever possible with the future Government sponsored offshore programmes.

Existing methods. Mooring loads due to waves, currents and winds are usually treated as independent variables. This approach may be adequate in some situations but will require detailed consideration of the interaction between waves and currents where the latter are significant. However, some special difficulties need to be evaluated further:

— low frequency drift forces due to wave action can cause large amplitude motions of the converters at resonance. Such dynamic effects may amplify the mooring forces considerably. There are both theoretical and experimental difficulties in estimating these low frequency forces;
— the estimation of drift forces is unreliable in random seas, and approximate methods have been developed to extrapolate regular wave data to random sea conditions. There is doubt about such a procedure, however, and work is in progress at the National Maritime Institute (NMI) to clarify the situation (as part of an OETB programme);
— the calculation of the surge response of ships in seas of irregular head is unlikely to be adequate when applied to wave energy converters.

Methods needing further development.
Whilst existing theoretical models, assuming linear waves and response, will probably be satisfactory for moderate operational conditions, they are unlikely to be applicable to extreme survival conditions. Continuing development is needed to increase the understanding and quantification of the phenomena involved.

The relationship between peak and rms drift force (also low-frequency response and mooring loads) is different from that for wave heights. Analytical work is needed to enable designers to make more reliable estimates of long-term statistics of mooring loads.

Instabilities due to non-linear terms in the equations of motion of the converters indicate that large mooring loads in certain sea states can occur. An investigation is needed into possible ways in which these effects may arise and whether the proposed mooring arrangements will be affected.

Non-linear analysis in the time domain needs to be studied to determine its importance relative to linear analysis when examining large amplitude responses associated with instability or extreme sea states.

A programme at NMI on the drift forces on ships in regular waves could be extended to multi-body systems and used in conjunction with existing approximate formulae for random sea forces to provide designers with some conception of the magnitude of the dynamic problem.

A tentative conclusion on mathematical modelling is that moderate sea conditions can be modelled effectively, and the analysis will be particularly useful in determining long-term fatigue effects on components. However, until such time as the mathematical models are capable of dealing with extreme and survival conditions model tank tests are likely to be the most satisfactory method of providing design data.

Tank model testing

Tank testing has been a vital component in the development carried out by the device teams. Until recently most of this work has concentrated on the hydrodynamic performance of the converter models and the moorings have been artificial in the sense that the models have been either fixed rigidly or restrained by moorings of arbitrary stiffness.

Testing which is designed to determine the mooring forces is now under way, although the emphasis differs in each of the device teams. The Cockerell raft system has special problems which arise from the proposal to use mooring lines to separate and control adjacent raft units: there will be therefore a dynamic interaction between these units in addition to the interaction with the seabed moorings. One unresolved question on all floating converters is whether each unit should 'weather vane' the seas or whether it should be moored with a minimum alignment capability. The latter would reduce the problem of orienting the power cables between the converters and the sea floor, but many interrelated factors will have to be considered before final decisions are made on this point.

Preliminary predictions of full-scale, mean mooring forces from model tests have ranged from 1 to 2 tonf/m in the case of the Salter duck string to up to 10 tonf/m for the Oscillating Water Column and the Cockerell raft.

Ideally the force on the converter should not increase in proportion to an increasing sea state. There is evidence to suggest that this situation may be achieved by a duck if it is allowed to flip on its back in response to the largest waves. The other teams have also recognised this ideal and will be working on methods to keep mooring forces to a minimum in extreme conditions.

It is regarded as of the utmost importance that tank testing methods used are representative of real sea states—a definition that has yet to be finalised—and that known non-linearities in the behaviour of bodies in wave systems and in response to mooring dynamics are modelled correctly.

Component integrity

To ensure both the initial and long-term integrity of large-scale wave energy converters at sea, it is visualised that extensive land-based component testing under simulated loading would need to be carried out. This could range from relatively straightforward fatigue testing of man-made fibre ropes for example (already under way at small scale) to complex multi-directional fatigue loading of a structure connecting the cable to the converter.

Accurate and extensive surveying of the sea floor at the chosen locations will be necessary to ensure the correct choice of anchor. Whilst it is not thought that there will be insurmountable technical problems in providing anchorages, research and development may be needed to prove anchors at the large sizes envisaged.

To ensure their ultimate integrity, the first large-scale converters to go into real sea conditions are likely to be moored conventionally—or unconventionally—by over-designed mooring systems making use of second or even third stage fall-back systems wherever possible. The limit to this approach of course is the need to allow the converter to behave in a 'normal' manner during all modes of operation.

Based upon the present procedures laid down by the certifying authorities, annual and quinquennial inspections would involve the considerable use of techniques such as underwater television and the deployment of divers. Samples of the mooring system would have to be recovered for inspection on shore every five years or so. Much further work will be necessary to determine the extent to which inspection and maintenance can be reduced—especially the extensive use of divers. However, in the early stages of the programme, including prototype testing, inspection will need to be at frequent intervals until information on typical wear and degradation is built up and hence an optimisation of capital and maintenance costs achieved. Full monitoring of mooring loads, device involvement and environmental forces with mooring designs evolved during the R and D period should permit the design of safe and economic moorings.

5.4 Concluding remarks

In summary, the development of wave energy converters can proceed unhindered for several years using over-designed mooring systems based on existing knowledge to ensure adequate reliability of the models or prototypes under test in the open sea. However, the ultimate feasibility, technical and economic, of all designs of floating converter will depend upon extensive further work of which a first outline is given above: the present state of mooring technology is inadequate. Obviously any design of converter which involves building upon the seabed, such as the Rectifier, can eliminate the mooring problems completely, but may pose other problems of similar or even greater magnitude.

6 Energy conversion and transmission

The engineering development of the various designs of converter by the Device Teams has been supported by general work in the area of energy conversion and transmission sponsored and coordinated by Technical Advisory Group 6 of the Wave Energy Steering Committee (see Appendix 3). Before discussing that work it will be illuminating by way of introduction to consider the characteristics of waves as the energy source, building upon the basic material in Chapter 2.

The waves of principal interest are characterised by random heights distributed over a range of frequencies of the order of 0.1 Hz: a typical spectrum is shown in Fig 6.1(a). Wave energy converters will respond to these waves with an efficiency which varies with frequency, as illustrated in Fig 6.2, and which is determined largely by the characteristics of the applied load. An ideal load for most designs of converter would be one which generates forces proportional to the velocity of the converter motion (a 'linear' system); a typical converter response with such a load is shown in Fig 6.1(b) and the power associated with this response is detailed in Fig 6.1(c).

It becomes immediately obvious that:

— the energy conversion system must be able to handle large short-term variations in the instantaneous power level
— the peak power level can be many times greater than the average power level
— the primary power output is not in a form that can be handled conveniently.

Superimposed upon the short-term variation of the available power are longer term effects resulting from day-to-day variations in the sea state and seasonal variations in weather patterns. The seasonal cycle at OWS India, for example, varies between an

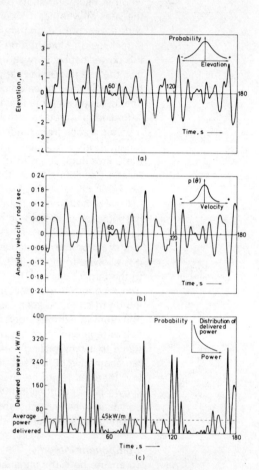

Fig 6.1 *Derivation of power spectrum from a typical spectrum of wave heights*

(a) A typical train of waves (Pierson-Moskowitz spectrum; T_z = 9 s; H_s = 4 m; average power = 80 kW/m)

(b) Angular velocity of device response

(c) Power delivered to first-stage conversion system (average power in incident waves = 80 kW/m)

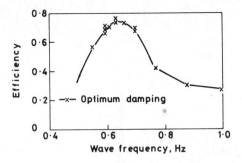

Fig 6.2 *Typical efficiency variation for a model tested in a narrow laboratory tank*

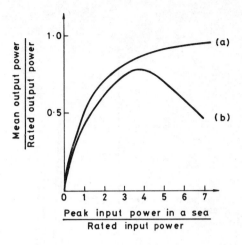

Fig 6.3 *Effect of peak equipment rating on the delivered power*

average power available of ≈180 kW/m in the winter and ≈30 kW/m in the summer. The implications of these variations for the electricity supply system are discussed in Chapter 7.

The randomness of the wave energy availability prevents the establishment of simple physical models for use in wave power studies and it is usually more convenient to work in spectral and statistical terms. In the insets to Figs 6.1a and 6.1b are representative spectral density functions for both the incoming waves and the resulting converter motion. In each case the distribution of elevation for the wave or the response velocity $p(\theta)$ of the active converter element is approximately Gaussian with a mean of zero. The power output is proportional to the velocity squared in a linear system and the instantaneous power is approximately as shown in the inset to Fig 6.1(c).

Thus there are a significant number of occurrences where the power output is much greater than the mean and, if the power take-off equipment is not to be rated for excessive power levels, some proportion of the available power has to be shed in a controlled fashion.

The manner in which this shedding is achieved depends upon the particular design of converter and the method of control. Where it is possible to continue to generate at the design peak even when it is exceeded, the overall conversion is as shown in Fig 6.3, curve (a). Some protection methods, a slipping clutch for example, may cause the output to fall to zero in overload conditions with a consequent lower overall efficiency. In practice a converter will exhibit variations in conversion efficiency over the whole range of applied velocity/power levels. This is illustrated in Fig 6.4 for a system with an efficiency curve which peaks at some intermediate power rating and is low at relatively low power levels. The overall conversion efficiency of this system as a function of peak rating of the conversion equipment is as curve (b) in Fig 6.3.

From the foregoing it is clear that the precise characteristics of the power conversion equipment, and the associated control and protection arrangement, have a marked effect on the power conversion capability. Furthermore, such considerations are an important contribution to controlling the range of power levels which the bulk transmission system has to handle. In this respect it is desirable both from a transmission and utilisation point of view for the short and medium time scale variations in the output power of the converters to be kept to a minimum.

This, to some extent, may be achieved by the interconnection of several converters into large integral units, which in any case is essential for bulk (>100 MW) transmission of the energy. Here the diversity of the individual converter outputs, resulting from the directional properties of the waves, leads to a smoothing effect when they are combined. The degree of smoothing is a function of the overall dimensions of the integrated unit when compared with the wave- and crest-lengths of the wave spectra. Where any envisaged arrangement of converters does not provide an adequately smoothed output, it becomes necessary to

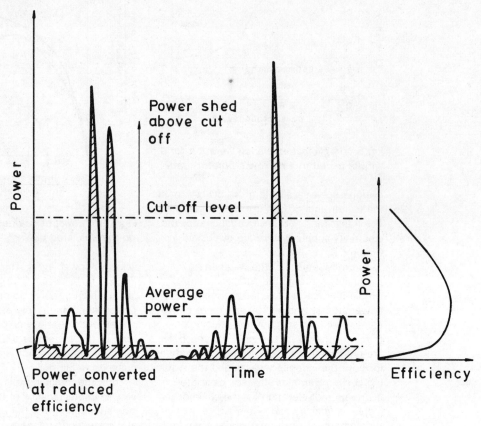

Fig 6.4 *Interaction of the conversion characteristic and the power extracted by the device*

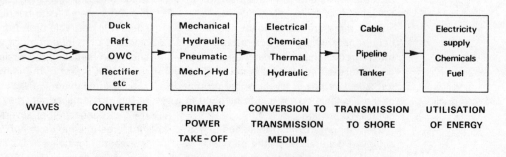

Fig 6.5 *Basic flow diagram*

include some form of short term storage within the energy conversion system (see also Chapter 7). Since smoothing through diversity, or storage, results in equipment ratings significantly less than the peak it is important that it should be achieved as early in the conversion chain as possible.

6.1 Conversion and transmission systems

The basic flow diagram detailing the essential elements of the energy conversion and transmission system is shown in Fig 6.5, and a more detailed display of the wide range of systems to be analysed and

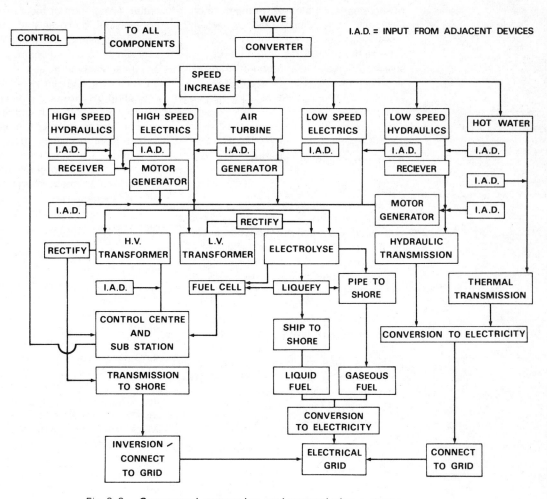

Fig 6.6 *Conceptual conversion and transmission systems*

choices to be made is given in Fig 6.6.

The primary output of each converter is quite different but can be characterised in one of three basic categories. Designs such as the Duck or the Raft, in which power is derived from the relative motion of large solid structures, are characterised by a low velocity-high torque primary power take-off system. Such a system normally has to be interfaced with a secondary energy conversion system before the energy can be utilised.

Oscillating Water Column designs operate basically in a similar manner but one of the structures is replaced by a water column which acts as a large reciprocating piston delivering air to an air turbine. In this case, direct conversion is relatively straightforward although the characteristics of the fluctuating air flow are somewhat unusual.

A third primary power take-off option forms an integral part of the Rectifier. This utilises two reservoirs and a system of gates by means of which waves charge a high level reservoir and empty a low level reservoir on alternate half cycles. Power is generated by a low-head water turbine positioned between the two reservoirs.

Although the primary power outputs are seen to be device-specific, there are several possibilities for the later stages of conversion and transmission which are common to each. Here the energy carrier can be:

—electricity;
—pressurised fluid;
—chemicals;
—heat.

These energy carriers could be transmitted to shore either by a cable, in the case of electricity, or by pipeline (and/or possibly tanker ships) for the other options.

Since the primary take-off is device-specific whereas the subsequent transmission is of a more general nature, it is convenient to treat these as separate subsystems.

6.2 Primary power take-off

Duck and Raft designs

A mechanical connection between the device and the first stage converter is thought to be essential. Such a connection needs to be able to transmit the peak torques of a device over long periods with reversing loads in a marine environment. For example, a raft transferring 100 kW/m at one hinge has a torque loading of 1 MNm per metre width of device. Wide variations in device velocity have also to be accommodated.

The main options for this type of duty are chains, belts, friction drives including tyres, gears and crank and connecting rod assemblies. 'Order of magnitude' capabilities of these options are listed below:

	per m width of device
Cranks/connecting rod	3 MNm
Gears (per pinion)	0.7 MNm
Cams	0.7 MNm
Roller chains	0.1 MNm
Belts	0.01 MNm
Friction drives	Small

Cranks and connecting rods are seen to be capable of transferring the torque but require most of the available width of the device. Pressure lubricated journal bearings are suitable for the crank pin with rolling element bearings being a possible alternative. Gear drives are also capable of transmitting the required torque in both the Duck and the Raft, although there must be an emphasis on good design which includes not only the gear form but also close tolerances on gear spacing and the need for a good lubrication system. In this context, sea-water lubricated gears with a significant working life are not thought to be an economical proposition.

Probably the simplest conceptual conversion scheme for use on wave energy devices is a directly-coupled electrical generator. Generation at the device speed is not possible, however, and some form of speed-increasing system becomes necessary. Such an arrangement requires the electrical machine to be capable of meeting an onerous overspeed requirement, thus restricting the size of the machines to ≈ 100 kW/pinion. Although large numbers of small machines offer the prospect of greater plant availability, their outputs are more difficult to combine for bulk transmission. Direct coupling of electrical machines is also likely to impose large inertial loads on the mechanism and backlash becomes a major consideration. For these reasons direct generation is not considered a feasible proposition.

Alternative options for use as the primary load for the Duck and Raft are low speed hydraulic systems. The mechanical coupling of hydraulic rams to each of these devices is feasible with both cam or crank arrangements provided a sufficient number of rams is installed to share the load. Hydraulic rams are particularly suited to a marine environment because of their simplicity and ability to handle a low speed reciprocating input. Further advantages arise because of their known reliability and the ease with which several ram outputs can be combined at a central storage/conversion point removed from the sea environment. Hydraulic storage allows some smoothing of the power input to the second stage conversion arrangement, and the hydraulic interface can be used for control purposes to provide an optimised load characteristic to the device. In this instance control can be exercised by varying the number of rams in circuit as a function of the state of the sea. Several small rams properly matched in number to the immediate sea state can give a substantial improvement in overall conversion efficiency compared with a single large ram.

As with the mechanical option, hydraulic speed-increasing systems are possible. A pump/motor combination is capable of speed variation over a wide range limited only by the availability of suitable motors and pumps. The hydraulic approach is less efficient than would be tolerable in a gear designed for a long low-maintenance life, but the difference may not be significant. Importantly, both gearing and directly coupled hydraulic speed-changing will give rise to a motion at the output which ranges from full speed to zero in each wave period and will transmit the final generator load characteristics back to the device without modification. The hydraulic approach would in general give rise to a unidirectional motor drive and be more amenable to large speed increases, both of which could prove to be advantageous.

It is therefore not surprising that hydraulic equipment forms some part of the primary conversion function of the favoured systems which have evolved for the Duck and the Raft. One of the primary power take-off systems currently under consideration for the Raft in mid-1978 was a gear drive with some speed increase to a cranked ram arrangement. The working fluid would be filtered seawater pumped through a low pressure, high volumetric flow hydraulic main, incorporating some accumulator storage, to a reaction turbine in the base of the centre raft. In general, such a system would be expected to avoid the cavitation problems associated with higher flow velocities but further study of cavitation and corrosion effects will be necessary for wave power devices utilising sea water.

Thinking in the Device Team in 1978 on the primary power take-off for the Duck involved either a geared or belt drive arrangement connected to hydraulic pumps feeding a closed circuit, high pressure oil main. Several second stage conversion options could be incorporated in such an arrangement and the favoured options were directly driven hydraulic turbines or a pressure exchange unit converting the oil flow to sea-water flow to drive a turbine.

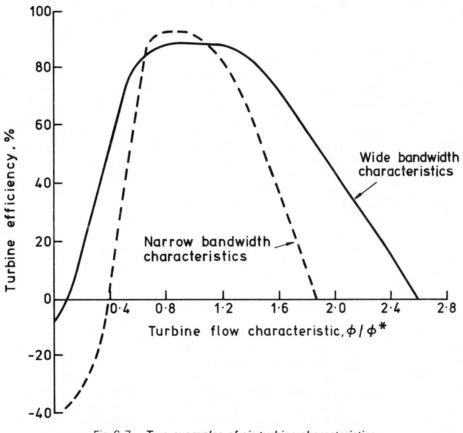

Fig 6.7   Two examples of air turbine characteristics

$$\phi/\phi^* = \frac{80Q(\theta)}{D^3\pi\omega_T}$$

where $Q(\theta)$ = instantaneous air turbine flow rate
  ω_T = angular frequency of turbine shaft
  D = diameter of impeller

Oscillating Water Column design

The overall conversion concept, involving a fluctuating air flow through an air turbine is simpler and is closer to final definition than those for the Duck and the Raft. The major area for quantification lies in the interaction between the fluctuating air flow and the turbine/generator characteristics.

The efficiency of the total power conversion, sea-wave to electrical energy, depends on the device efficiency, turbine efficiency and generator efficiency. The device efficiency is determined essentially by the loading characteristic reflected into the device, while the turbine efficiency is determined by the specific turbine characteristic, the efficiency of the air flow rectification arrangement, energy storage and generator control philosophy adopted for the defined sea-state. For example, the behaviour of conversion systems with two different turbine characteristics as detailed in Fig 6.7 can be simulated for particular sea conditions, to assess the effect of various generator inertia constants. Inertial energy storage is essential for smoothing the large fluctuations in available power, and for any given set of conditions there exists an optimum inertia such that it is large enough to avoid repeated instances where the turbine/generator unit might overspeed and yet not so large that the inability to accelerate results in an excessive loss of turbine efficiency.

The simulation for the two different inertias is shown in Figs 6.8 and 6.9 for an idealised monochromatic sea with a wave gust. The large increase in air flow rate produced by the wave gust (Fig 6.8c) is seen to increase the speed of the turbine with the wide bandwidth (Fig 6.8a) so that it is able to maintain full output power, whilst the speed of the turbine with the narrow bandwidth characteristic falls away rapidly.

During the subsequent low wave period, energy is extracted at similar efficiencies for both inertia cases but with H = 5s (the inertial time constant) the electrical power at times drops to zero as there is neither enough stored energy nor converted energy to feed the demand. With H = 15s, although there is a substantial drop in electrical output power, the energy stored in the inertia from the previous monochromatic period is sufficient to supply a reduced electrical output. In the ensuing monochromatic period the turbine again gradually settles into its 'steady' operating regime in a time dependent on inertia, with the smaller inertia machine exhibiting the most rapid response.

Thus, in cases where the turbine characteristic exhibits a narrow bandwidth, in order to maintain acceptable turbine conversion efficiency over a large range of flows *low* inertia and a large speed variation are required which may itself invoke the overspeed criterion (or create mechanical design problems in the turbine) and/or produce power outputs with large variations depending on the generating characteristic selected. For operation in low waves during the gusting period it is advantageous to have substantial stored energy capacity to prevent large dips in output power. In this respect the *higher* operational inertias allowed by a wide bandwidth characteristic are preferred.

Obviously studies of this nature have to be extended to real sea conditions but the above simulation does provide an insight into the factors which influence this conversion option. A further consideration involves the choice between a conventional uni-flow air turbine and a self-rectifying turbine which does not require air flow rectifying arrangements. At least two self-rectifying air turbine concepts exist and appear promising but significant research and development will be necessary before the attraction of increased reliability, and possibly efficiency, as a result of dispensing with air flow rectification equipment outweighs the known efficiency and almost linear head-flow load characteristic of a radial inflow turbine.

Rectifier design

One device which does contain substantial inbuilt storage is the Rectifier. The primary power take-off system is a low-head turbine. The expected head is 1–3 m and this is somewhat lower than normally encountered during water turbine usage but is not so far removed as to present insuperable problems. The Rectifier can also be a seabed-mounted device and as such will probably be located closer in to shore than the other devices. This factor, together with the inbuilt storage of the Rectifier offers the prospect of synchronously coupling the output of the unit with the National Grid.

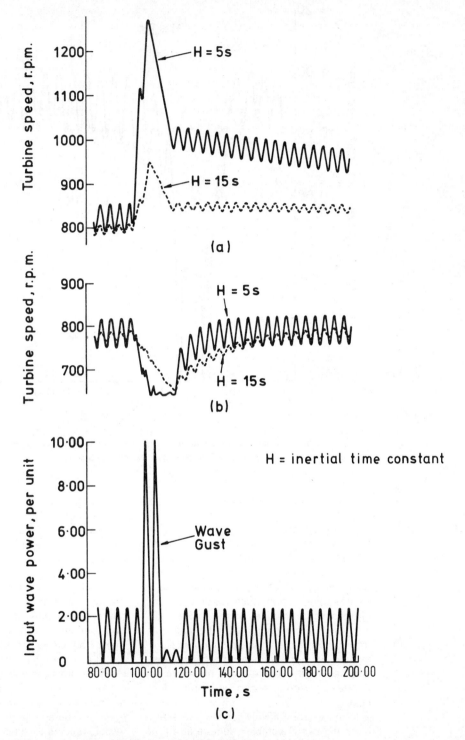

Fig 6.8 *Interaction of inertia and turbine characteristics*
(a) Wide bandwidth characteristic (output current = 1.0 per unit)
(b) Narrow bandwidth characteristic (output current = 1.0 per unit)
(c) Ideal 'gusting' sea state

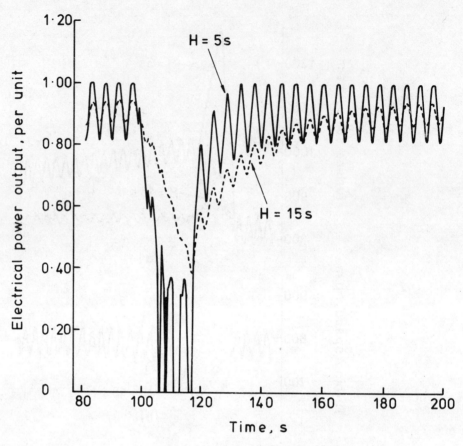

Fig 6.9 *Variation of electrical power output (and terminal voltage) Current = 1 per unit (narrow bandwidth characteristics)*

6.3 Conversion and transmission

Electrical option
The basic philosophy of any electrical conversion and transmission system would be to group progressively the outputs from the wave energy converters into a main high voltage transmission cable. Since the cost of transmission reduces as the voltage is increased, it is economically advantageous to achieve the maximum grouping as early as possible in the transmission chain. This also has the advantage that a minimum of the transmission system need be rated for peaks.

In most cases, however, the wave energy converter is not a stable structure at a fixed position in space, and any power take-off must be via a flexible cable. At the present time the upper limit based on insulation for flexible cabling is 22 kV, and the maximum rating of the order of 10 MW is too small to take the average power from an array of sufficient dimension to give a smooth output: except when substantial hydraulic smoothing or averaging is provided the cable rating must take peaks into account. In the extreme, 10 MW would perhaps serve only 100 m of device were it not for the fact that transmission cable is more tolerant than other equipment of short duration overloads. Without significant development work on cables it would not be possible to arrange for all the equipment up to main transmission voltages to be mounted on the converter because the power could not subsequently be taken off. There must, therefore, be a large number of low power flexible feeders to substations on offshore platforms or the sea bed.

For generation on board each converter unit there is a choice of either AC alternators or DC generators. DC machines can in principle be grouped either in series or in parallel. Apart from other difficulties parallel grouping restricts the total voltage to about 1 kV because of commutation considerations. If connected in series, DC machines with individual quick-response excitation control could provide whatever speed-torque characteristic may be required. However, to maintain the insulation levels between armature and frame which would be necessary at the ends of the series string could be difficult if machines with brushes were used. Electronically commutated machines might well be feasible for this purpose. These machines shade into true AC machines. Synchronous transmission using AC generators is impractical due to the wide variations in speed of the converter units in the constantly changing wave conditions. (The Rectifier is a possible exception to this.)

The favoured system is shown schematically in Fig 6.10. Conventional synchronous generators operate into individual diode rectifiers, which are series connected, in order to minimise interaction problems. A shore based inverter station connects the DC to the national grid and operates in a constant current mode. Overall transmission voltage is set by the summation of individual rectifier outputs with the constant current control at the inverter determining power transfers. Consequently each generator will operate at constant armature current, varying voltage; with voltage and power variations at individual generators being determined by the generator operating philosophy and energy storage capability.

The scheme relies on the assumption that a line of devices of reasonable length, up to say a combined output of 200 MW, provides a substantially smooth output, and a negligible amount of smoothing between the alternator and wave energy converter would be required. If smoothing is found to be necessary it should be possible to achieve this by the use of flywheels on the alternator rotor shafts. Without field control, the natural torque/speed curve for the alternator is not the ideal linear relationship, but it would be possible to achieve a very close approximation to the ideal characteristic by controlling the excitation of the machine.

Figure 6.11 shows the current state of the British grid network and an indication of the preferred general locations for siting wave energy stations. Wave power systems located off the SW and NE coasts would be sufficiently close to the present grid network to avoid significant grid reinforcement. This is not the case for a wave power system located off the Outer Hebrides and, although the first 2 GW of installed capacity would be absorbed to meet the increasing Scottish electricity demand, a major grid reinforcement would be required to accommodate the wave power system if it reached its full potential. Even so, electrical conversion and transmission is recognised as probably the cheapest option under consideration, with the major costs centred on the rectifier/inverter equipment and its associated housing. The electrical costs are not particularly sensitive to the cost of the high voltage DC bulk transmission cable, which for a system located 20 km off-shore represents 2–3 per cent of the total cost per kW of installed capacity—although the need to protect the cables with trenching would substantially affect this. Thus the cost of the electrical option may not be affected significantly by the exact siting of the units or the depth of water in which they operate (unless sub-sea transformer/rectifier units are employed).

Hydraulic option

The hydraulic option is a modified version of the electrical option with the generation equipment sited on shore and connected to the wave energy converter complex by an under-sea hydraulic main. The removal of the electrical equipment from a potentially hostile environment offers the benefit of greater reliability, and the electrical interconnection of large shore units ($\approx$80 MW) with the grid is much simpler than in the 'electrical' scheme.

Two transmission fluids are possible:

—filtered sea water;
—conventional hydraulic oils.

Sea water has many attractions over oil as a means of transmitting power; most notably it requires only one pipe circuit. Thus the combination of a shorter pumping distance and lower viscosity makes high pressure water transmission more efficient than the oil alternative. An analysis of transmission efficiencies, based on a converter to shore transmission distance of 24 km using 1.2 m diameter pipes at a pressure of 15 MN/m^2 is shown in Fig 6.12.

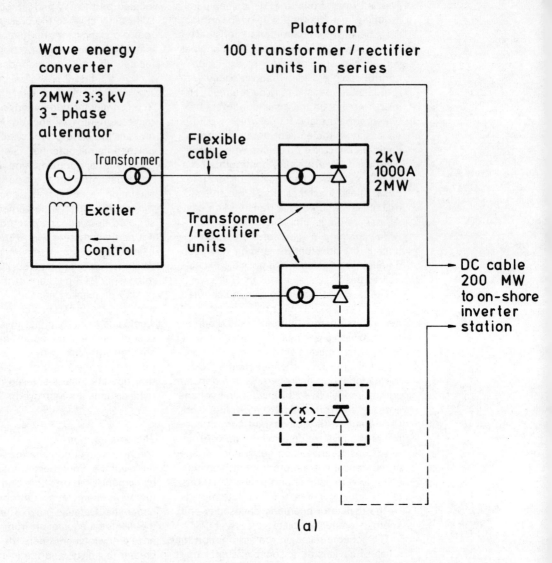
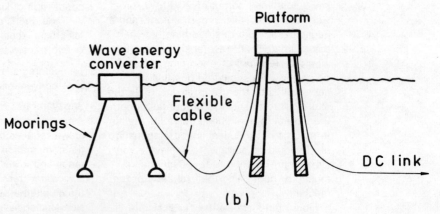

Fig 6.10 *Preferred electrical scheme*
(a) schematic
(b) pictorial

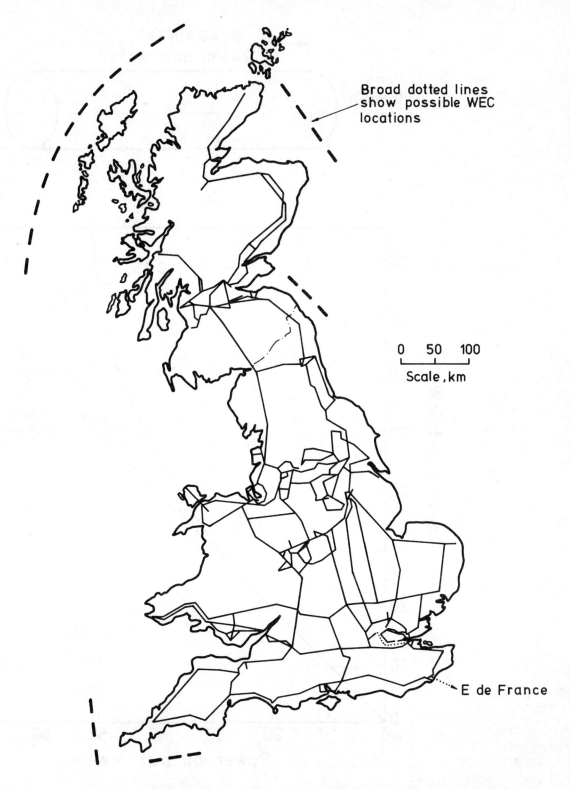

Fig 6.11 *Major routes of the 400/275 kV transmission system in Great Britain*

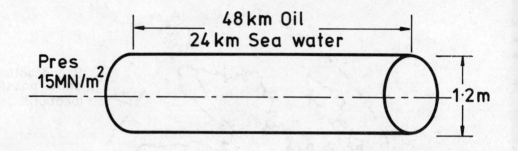

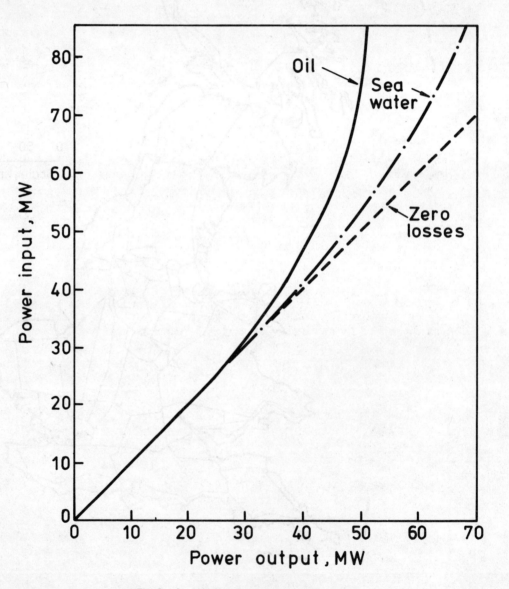

Fig 6.12 *Hydraulic power transmission*

The 1.2 m diameter pipes, a size which will soon be available for pressure up to 15 MN/m², would need to be concreted over for protection and ballast but even so some problems may be encountered in shallow water (< 50 m) due to cross currents and shipping. Trenching would be essential to give the pipe continuous support and this could prove difficult and expensive in regions where the sea bed is mainly rock.

Connection of the converters to a seabed hydraulic main would be by rigid pipework with a 'trombone' or articulated riser to compensate for heave. Limited lateral motion could be accommodated by ball-jointed connectors. If converter movements prove to be excessive, a semi-submersible structure could be used for termination of the hydraulic main with flexible pipework, which is now developed to an advanced stage, to the surface. Furthermore, the hydraulic main could be designed to act as a hydraulic accumulator of very large capacity, thus providing a relatively smooth delivery of power to the shore-based conversion equipment. The major disadvantage of such a system is the cost of manufacture and installation of the seabed pipeline equipment. Initial estimates suggest that hydraulic transmission is at least two to three times as costly as electrical transmission whilst it only, in effect, replaces the electrical inverter/rectifier equipment and DC cable link of the electrical option.

Chemical energy carriers

The utilisation of wave power to produce chemical products, which can be transmitted to shore by pipeline or tanker, offers the avantage of freedom in the siting of the converters and the energy landing sites. Most chemicals can also be stored cheaply and easily for long periods and may therefore eliminate the problem of mismatch between the wave energy supply and the generating system demand which arises when the desired output is electricity.

To date, all of the chemical products which have been considered for wave power systems are based on the production of hydrogen by the electrolysis of water. These are:

—hydrogen;
—ammonia;
—methanol;
—methane;
—gasoline.

Ammonia is produced by the catalytic combination of hydrogen- and nitrogen-using technology which is well established on a commercial scale. Apart from water, nitrogen is the most abundant material available to the wavepower-chemical complex, constituting some 79 per cent by volume of the atmosphere. Ammonia is a basic feedstock for several large scale chemical production processes, especially in the fertiliser industry. Ammonia-based fertilisers comprise a vital commodity in our economy, for which the demand is expected to almost double by the year 2000. That future level of demand translated into wave energy terms would require a continuous power output approaching 3 GW. It is conceivable, therefore, that the available wave energy could supply part or all of the basic energy input in the United Kingdom for ammonia-based fertilisers (which at present is supplied by natural gas).

Methanol, methane and gasoline are produced by various catalytic combinations of hydrogen and carbon oxides although here it should be recognised that it may be impossible to deliver carbon oxides to the chemical complex at an economic cost.

The electrolysis of water requires the first stage (medium voltage) of the electrical conversion option but eliminates the problems involved in the electrical series connection of many converter outputs for bulk transmission. A possible production route would be to use the rectified output of directly-coupled AC machines to electrolyse water and produce hydrogen at 30 bar. Although an electrolyser is not a purely resistive load, it is possible to design a close approximation to velocity-proportional damping with a synchronous alternator/electrolyser combination, which is closer to the believed ideal load characteristic than most of the other systems considered.

The direct electrolysis of sea water is not a practical proposition for wave power systems and desalination on board the converter would be an essential first step. A schematic flow diagram for hydrogen production using current technology is shown in Fig 6.13a.

Advanced electrolysers of greatly improved efficiency and reduced capital cost are being developed in energy programmes

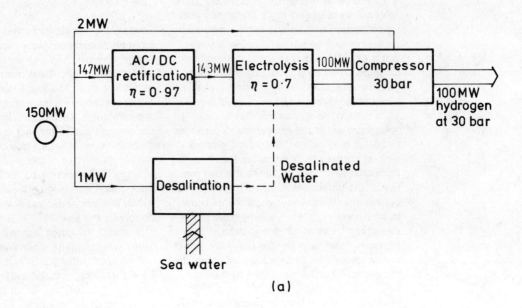

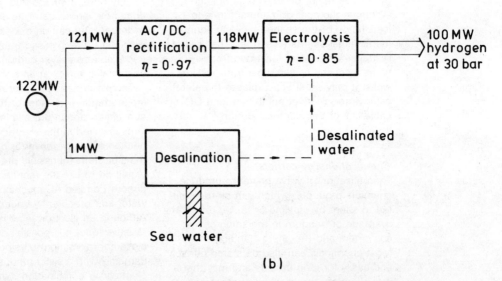

Fig 6.13 *Schematic flow diagram for hydrogen production*
(a) using existing technology
(b) using advanced electrolysis techniques

promoted by the EEC and the International Energy Agency. The schematic flow diagram for such a system is shown in Fig 6.13b, based on an electrolyser efficiency of 85 per cent, representing operation of the system to minimise the capital cost of installation. Operation of the advanced electrolysers with almost 100 per cent efficiencies should also be possible, but at an increased capital cost.

Several basic configurations are possible for siting electrolysis and chemical plant and the transmission of chemicals or energy between them. Schematic block diagrams of these are shown in Fig 6.14. They involve:

(a) siting the whole electrolysis/chemical complex at sea and transmitting the chemical product to shore;
(b) siting electrolysers at sea and transmitting hydrogen to shore-based chemical plant;
(c) siting the whole electrolysis/chemical complex ashore and transmitting energy to it.

With the whole chemical complex at sea, mounted on a stable platform serving several converter arrays, the chemical products could be transmitted to shore either by undersea pipeline or tanker ships.

The technology for laying and operating undersea pipelines exists and is developing continuously. It is being extended to deeper waters and more hostile conditions as the

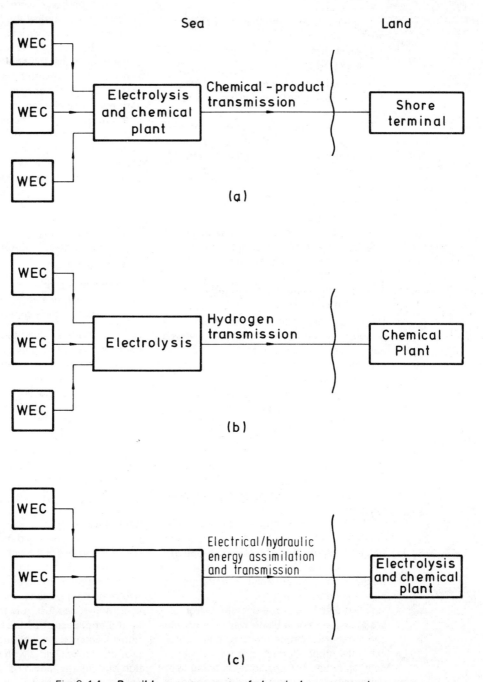

Fig 6.14 *Possible arrangements of chemical energy carrier systems*

need arises, as for example in the North Sea. The main elements of cost for a pipeline transmission system are:

—materials for the pipeline. As the depth of water increases thicker-walled pipes must be used to withstand greater hydrostatic pressure and the large stresses occurring during the laying operation;
—pipe laying;
—marine riser in the case of a surface installation;
—pipeline compressor(s), both capital and power costs.

The total pipeline transmission costs are therefore dependent upon distance and water depth.

Transportation of liquid chemical products by ship is again known technology, even for those such as methane which must be transported in cryogenic vessels. The main costs are:

—capital cost of ships, including on-board storage tanks;
—facilities for product storage at the chemical plant. This method of transport is essentially a batch type of operation;
—ship operating costs, i.e. labour, fuel, maintenance.

Total costs for this form of transmission will obviously depend on the distances involved, but will be independent of the water depth.

If electrolysis only is performed on the converter (Fig 6.14b), hydrogen is the energy vector transmitted to shore. Hydrogen transmitted by undersea pipeline can be utilised either by a local chemical plant or be transmitted overland by pipeline to a more central location. For tankered hydrogen, liquefaction would take place at sea using a large tonnage central liquefaction plant, and the product would be landed at any port.

When all of the chemical plant is located on shore, the conversion and transmission system is identical to the electrical option with simply the large overland electrical transmission lines replaced by underground pipelines or ships.

A first estimate of the overall cost of producing ammonia using wave energy indicates that it would be more expensive than using the other alternative to natural gas—coal. However, the possibility is being kept under review as an insurance technology.

Thermal transmission option

The conversion of wave power to thermal energy in the form of heated fluid is a feasible system which has some clear advantages. The energy conversion can be either a friction brake arrangement or resistance heaters fed from a DC machine, the latter having the disadvantage of requiring an initial conversion to electricity.

If a friction brake were used, it could be controlled to optimise the load characteristic seen by the device and supply heated oil or water to an insulated storage vessel. Several such vessels would be towed to and from the converter array transporting pressurised water or high temperature oil with each vessel sized for, say, five days charge/discharge at the maximum continuous rating. The whole installation has the advantage of simplicity and reliability although the operation of such a system is liable to be manpower intensive and subject to weather-window difficulties in the winter months.

The over-riding disadvantage of this option is that the energy is not available in a convenient form. Heated fluid could be utilised in three ways:

—conversion to electricity;
—to provide power station pre-heat;
—district heating schemes.

Even at an annual load factor of 70 per cent the thermal energy could not be converted economically to electricity by vapour turbines and the overall conversion efficiency would probably be no more than 3 per cent. Similarly, the provision of power station pre-heat is also difficult to justify economically and, as in the case of district heating schemes, the preferred sites for wave energy arrays offer a very limited application in any case.

Summary

Of the energy carriers detailed above, the thermal and hydraulic options appear too costly and inefficient to be given further consideration at the present time. Of the two remaining options, the electrical route is attractive because it is thought likely to be the cheapest of all the conversion and transmission routes although it does not provide the flexibility, with respect to location and energy landing sites, of the chemical option. Both these alternatives therefore remain candidates for use in wave energy schemes and, although further studies may define more closely the likely costs and benefits of each scheme, the final selection of energy carrier might be determined simply by the desired location.

6.4 Future studies

There is a need for more basic work on the problems of primary power take-off. Consideration will also be given to the design of bearings for use in a marine environment and to the design of, and materials for use as, seals. Studies will continue in the general area of hydraulic components and on the modelling of proposed hydraulic schemes for converters.

Future work on the electrical option involves defining more closely the design and operating characteristics of the various technical stages within the scheme. At the generation end of the transmission chain the feasibility of operating small groups of alternators, or possibly induction generators, on a common local busbar needs to be considered. Transmission of the AC power from the converter to the transformer/rectifier arrangement requires a cable capable of withstanding continual flexing with a realistic fatigue life, both mechanical and electrical. Work to assess the suitability of current cable designs and materials for this duty is to be undertaken.

The identification of the control problems of the electrical components within the conversion/transmission chain is another important field of on-going activity and this will be coupled closely with simulation studies of the behaviour of the various systems, including air and water turbine options, under 'real sea' conditions. Work in the immediate future on the chemical carrier option will be confined to monitoring the progress of electrolyser technology and the development of pipeline technology within the North Sea industry.

Finally, analytical work by the consulting engineers has pointed out that for several designs of converter an undesirable feature of the complete power chain from the waves to the user of electricity is the large number of conversion stages involved—energy is lost at each stage and the effects are cumulative down the chain. This can affect adversely both the economics and the usable fraction of the total resource. Whilst some aspects will be associated with the basic design and in particular with the ability of the converters to absorb energy from a spread of wave directions, a major objective of future work in this area must be to reduce the number of stages and to improve the efficiency of those which remain. For designs which involve an air turbine, there is scope for considerable work to optimise the matching of the turbine operation to the variable input.

7 Electricity supply system aspects of wave energy

From the start of the wave energy research programme it has been evident that large structures, sometimes with complex energy conversion and transmission systems, are needed to capture and supply wave energy in a useful form. Consequently equipment capital costs will be appreciable and paying for this capital, quite apart from any running and maintenance charges, is expected to result in significant overall costs for the delivery of wave energy to consumers (as foreshadowed in the 1974 report of the Central Policy Review Staff).

In the previous chapter the generation of electricity supplied for the national grid system and the production of chemicals were identified as the most likely routes for exploiting wave energy. The present chapter discusses some of the questions that are raised if wave energy is to be exploited within the context of the electricity supply system. The more fundamental of these questions are:

—what is the nature of the energy output from a wave energy scheme and how might it be modified so that the best possible economic return is achieved?

—what are the essential attributes of a wave energy scheme, as seen from the electricity supply viewpoint, and in what ways would it influence decisions to install other plant, and once constructed how would a wave energy system affect the operation of conventional plant?

—after due allowance for the benefits offered by wave energy, what are the capital cost targets that will have to be met if wave power is to make an economic contribution to possible future generating systems?

—given that the future is highly uncertain, how might the role of wave energy alter between different views of the future?

Unfortunately, there can be no firm or final answers to these questions at present for a number of reasons: there is a paucity of basic wave data for quantifying the characteristics of the resource; the costs and conversion characteristics of the designs currently under investigation are not well established; the designs chosen for a future wave energy scheme may not be one of those currently under investigation; and the future itself is uncertain. Therefore, the approach to the questions posed has been essentially one of clarification, and in order to clarify it has been necessary to make some assumptions, based on such evidence as is available, as to the possible characteristics of a future wave energy scheme and the future electricity supply system. Even with a successful research and development programme it is unlikely that wave energy could be exploited commercially on a substantial scale much before the turn of the century. Thus the assumptions for the system studies are chosen to be relevant to situations that could arise in the early part of the next century.

The value of the system studies lies not so much in the strategic planning of a future electricity supply system—an exercise that would be premature at the present time—but as an aid to R and D planning. The studies are important for defining cost targets, indicating how designs can be modifed towards an optimum, helping to choose between designs and indicating where R and D effort needs to be concentrated.

7.1 System characteristics of wave energy

Before any detailed studies were undertaken it was clear that wave energy would differ from plant currently used in generating systems. The output from fossil, hydro and nuclear generating plant, which make up the present generating systems, are largely under the control of the operators and more or less energy can be supplied in response to changes in consumer demand. By contrast, wave energy output will be largely a function of sea-state and weather. New units of conventional and nuclear plant are installed by the electricity supply authorities in order to provide, at a certain supply security standard, sufficient capacity to meet forecast increases in peak demand or to replace plant due for retirement. It is uncertain whether any installed wave energy capacity would contribute towards peak demand. In addition to contributing towards peak demand, new units or plant which have relatively low running costs, such as nuclear power, can reduce the need to operate other plant with higher running costs, such as fossil plant, and can thus make a financial saving to the system. It is envisaged that wave energy could contribute to a generating system in this way.

In the following paragraphs the likely characteristics of wave energy schemes are examined first and then used to indicate the influence of wave energy on the installation and operation of other plant in the system.

Using data for H_s (significant wave height) and T_z (mean zero crossing period), as defined in Chapter 2, together with assumed characteristics for wave energy device efficiency, conversion and transmission efficiency and plant availability, the

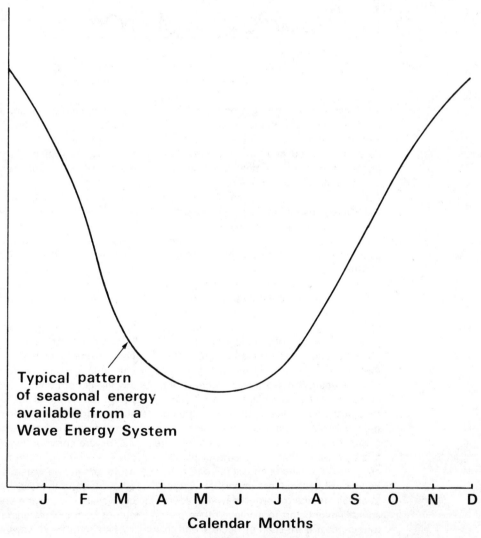

Fig 7.1 *Mean monthly wave energy supply*

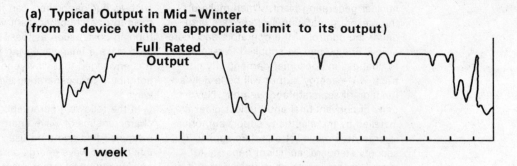

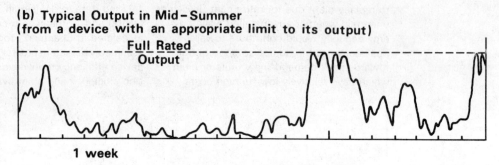

Fig 7.2 *Calculated output from a wave energy device*

electrical output that might be obtained from a wave energy station in a real sea can be estimated. Figs 7.1 and 7.2 indicate typical calculated monthly mean power levels and three-hourly mean power levels at a Hebridean coast location.

Figure 7.1 shows the expected result that there is more wave energy during the winter months than in summer. On a mean monthly basis, the pattern of wave energy supply broadly follows the pattern of electricity demand.

There are some advantages from having more energy available during the winter than the summer. In particular, in winter when average demand levels are higher, a greater proportion of the older and less efficient generating plant is operated. Therefore on average, small or modest quantities of wave energy could displace more of this higher operating cost plant. However, although the difference in efficiency between the smaller fossil plant and new, larger plants is significant at present, in the future it is anticipated that as the smaller plants are retired the difference will fall and this benefit will become less important. Large quantities of energy from renewable energy sources might start to displace other low running cost plant such as nuclear and this would occur most frequently when electricity demand levels are at their lowest in summer. Since displacing nuclear energy would bring less economic benefit to the system than displacing fossil fuels the greater availability of energy in winter has an advantage.

However, considerations of monthly mean energy levels is only a part of the story. Figures 7.2 and 7.3 indicate that there is a more serious problem in matching wave energy into electricity supply systems over periods of hours or days. Figure 7.2(a) shows that in the winter period, wave energy output at a given location typically would fall from the full device output to lower levels at irregular intervals lasting from a few hours to a number of days. It is not unusual, even in winter, for the wave energy at any location to fall to negligibly low levels from time to time. In the summer period full output is sustained for relatively short periods and low and zero outputs lasting for a week or more, as shown in Fig 7.2(b), are common. Figure 7.3 attempts to place wave energy output in the perspective of a total generating system and illustrates how other plant, in this case fossil plant, has to be available to provide energy when the output from a wave energy scheme falls.

The variability in wave energy supply could probably be reduced to some extent

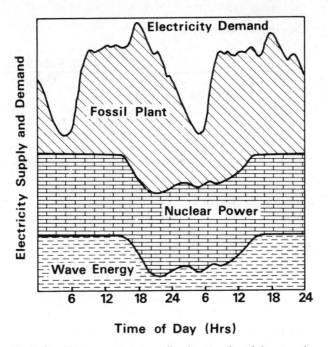

Fig 7.3 *Wave energy contributing to electricity supply. Typical pattern of electricity demand in winter showing how wave power might contribute to meeting supply requirements if the rated capacity of wave power plant represents about 20 per cent of the total capacity*

by combining the outputs from geographically well dispersed wave power locations: this is illustrated in Fig 7.4 which shows the averaged outputs from two locations about 250 km apart at the same instants in time. There appears to be some statistical independence in the times where periods of calmer seas arise between geographically displaced locations, and analysis of a limited amount of data for pairs of locations in the winter period from December to February suggests that the averaged output falls to very low levels relatively infrequently and the probability of a zero average output appears to be small. It is possible therefore, that some modest proportion of a wave energy system's output capacity could be relied upon to be available in winter. If further analysis proves this to be correct, wave energy might be considered to have some winter **firm power**. However, from such data as are available it seems unlikely that wave energy will have any meaningful firm power in summer.

Summarising these findings it is evident that wave power is very different from conventional plant. Its output is not controllable, other than by shedding energy which would otherwise be available. The energy generated has no particular relationship with daily or weekly electricity demand patterns. There is little certainty that full output would be available at times of high electricity demand.

Optimising a wave energy system
The variability in the output of wave energy can be reduced to some extent by lowering the rating of the conversion equipment on the device. Figure 7.5 shows the cumulative output from a wave power station as a proportion of the year. From this diagram it can be seen that the penalty of lowering the rating is loss of energy, but full output is sustained for a greater proportion of the time and the mean energy output level increases as a proportion of the rated capacity, ie the 'load factor' of the wave energy output is increased. Also to be taken into account in balancing the value of the lost energy against the improvements in variability are the facts that:

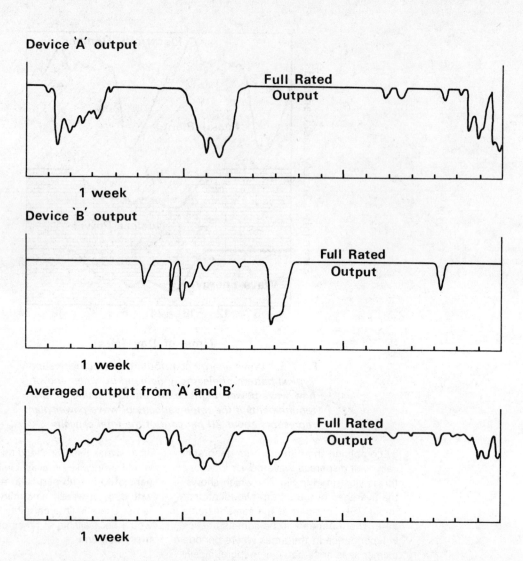

Fig 7.4 *Effect on variability of averaging the coincident outputs from two devices about 250 km apart in mid-winter*

Note: The wave energy output is assumed to be that from a typical device with an appropriate limit to its output.

— the on-board equipment costs fall as the equipment rating is reduced;
— lower ratings require a lower transmission capacity—ie a lower transmission cost;
— if there is any firm power, to a first approximation it will be independent of the equipment rating and therefore if the rating falls the *proportion* of the rated capacity that can be considered firm will rise.

Similar arguments can be applied to the choice of converter size since in most cases the larger the converter the greater the energy capture but the greater the capital cost.

By making assumptions about wave climate, the value of wave energy, the costs and performance of the converter and its on-board equipment, transmission costs and the value of firm power, it is possible to estimate the optimum size and rating and hence annual load factor for a wave energy scheme to achieve the best economic return. The optimum absolute size of the converter and its rating tends to be dependent on the wave climate that is assumed, but calculations indicate that the load

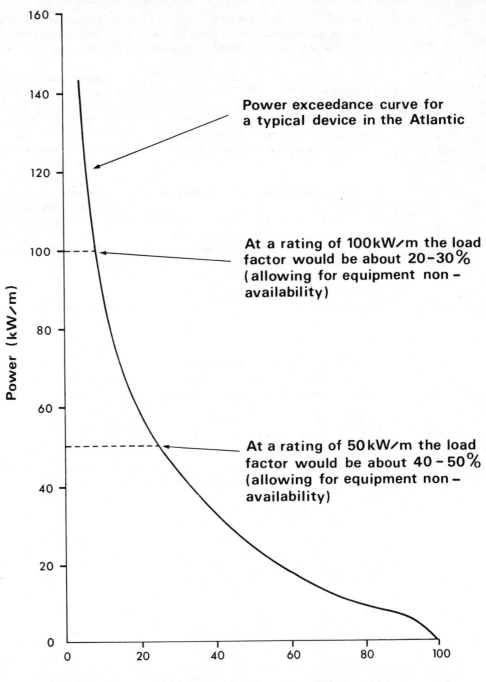

Fig 7.5 *Effect on load factor of reducing equipment rating*

factor is relatively independent of it. On the assumption that the costs of a wave energy scheme would be low enough for it to be economic in a future electricity supply system, it is probable that an annual load factor of about 45 per cent would be optimum (making due allowance for equipment non-availability due to breakdown and maintenance). As indicated earlier a well dispersed array of wave energy converters might have some winter firm power. There is little certainty as to how much firm power could be credited, but for a system operating at the optimum load factor up to 25 per cent of the rated capacity might be counted as firm on the basis of the limited amount of data available where the benefit of the uncertainty is given to wave energy. It will require the analysis of more data, particularly from locations closer to likely future wave energy station positions, before this or any other estimate can be considered reliable.

7.2 System implications of the characteristics

In the preceeding sections the lack of firm power and the variability of wave energy were noted as its distinguishing features. In this section the influence that these characteristics might have on the installation and operation of other plant in the system is briefly examined.

Firm power

A typical system which is made up of many units of conventional generating plant whose individual availabilities are relatively high but largely independent, has an expected overall availability of 85 per cent of the total output capacity at times of peak demand. It is reasonable therefore to assume that the firm power of each unit is about 85 per cent of its output capacity. As noted earlier, units of conventional and nuclear plant are generally installed to meet peak electricity demand at some accepted security of supply standard. However, the firm power that wave power can contribute is at best limited. Therefore installing wave power plant is not an alternative to installing conventional or nuclear plant for meeting peak demand—other plant will be required to compensate for the deficit in wave power's firmness.

In a system with a substantial wave energy capacity, the fact that wave energy is likely to have zero summer firm power, even if there is some winter firm power, could create some problems for the electricity supply industry in taking plant out of service for planned maintenance in the spring to autumn period. At present, after allowing for planned plant outages, the spare capacity available in the non-peak demand seasons is not much greater than in the peak demand period. Therefore a falling wave power firm capacity in the non-peak demand season could reduce the spare capacity available at that time sufficiently to require some re-scheduling of planned maintenance.

For these reasons the firm power benefits of a wave energy scheme are expected to be a relatively small part of its total value. Of far greater significance is the financial saving that can be made by using wave energy to displace the fuel that would otherwise be consumed in conventional power stations.

In a system with a large wave energy contribution one technical possibility for providing firm power in an electricity supply system might be storage. Storage plant brings firm power to a system whether or not there is any wave energy, and might be used where it could provide for a more economic match between generation and electricity demand. The test of whether a system with wave energy can benefit economically from storage is to determine the possible cost advantage from adding successive units of storage to a generating system in which wave energy is itself economic, and compare the outcome with the cost advantage of installing alternative types of plant.

The most usual form of storage is pumped hydro-power where the charge and discharge cycle covers 24–hour period. Given the lengthy periods when there may be little or no wave energy it is uncertain at present how much value daily cycle storage would have. Under circumstances where the combination of wave energy and energy from other low running cost plant, such as nuclear, exceeds demand (a situation which is most likely to occur during the night trough in demand) then the main options would be to:

—shed excess wave energy;
—reduce the output from nuclear plant;
—store the excess energy for release during periods of higher demand.

Storage might be the most attractive of these options, but only provided the circumstances occurred sufficiently frequently to justify the investment in storage plant. Storage plant with longer charge/discharge cycles covering several days would match better with the characteristics of wave energy supply and might therefore yield somewhat higher benefits to a system with wave power, but the cost of this storage is likely to be substantially greater than for daily cycle storage and whether or not there could be a net benefit is less certain. Therefore, although storage might be used to compensate for some part of the lack of firmness exhibited by wave energy schemes, it seems likely that unless some form of low cost storage can be developed most of the firm power deficit will be derived more economically from other conventional types of plant.

Variability of wave energy
From the viewpoint of the generating system the important timescales for considering energy variability are (i) seconds to minutes, (ii) minutes to hours and (iii) longer periods. Variability over periods longer than about a day can be accommodated with little additional economic penalty to the generating system. Allowing for this variability is essentially a matter of ensuring that adequate firm power is available to meet demands, and the penalties involved in this, as far as wave energy is concerned, have been covered in the preceeding discussions.

Variability over seconds to minutes. Fluctuations on the seconds to minutes timescale occur within the system at present as the result of unpredictable changes in demand and failures in generating plant. These fluctuations would cause the supply frequency to change significantly (outside accepted limits) if they were uncorrected. The necessary correction is usually provided by thermal plant operating below full output which can respond rapidly to changes in the system frequency. When completed the pumped hydro plant at Dinorwic, with its fast response time, will also make a valuable contribution to the regulation of system frequency. Wave energy might increase these seconds to minutes fluctuations and would therefore increase the requirement for frequency regulation capacity, with some modest consequential economic penalty. However, it is also possible that output smoothing in the form of on-board hydraulic systems (see Chapter 6) could reduce the fluctuations created by the wave energy scheme to a negligible level. Some designs, such as the Rectifier, have substantial in-built storage which would afford a high level of protection against short-term output fluctuations.

Variability over minutes to hours. Any significant variability on the minutes to hours timescale from a wave energy scheme would increase the need for flexibility in the operation of the remaining plant if a match between generation and electricity demand is to be maintained. It is not practical to start-up and shut down large thermal plants for periods less than about 6–10 hours and even this involves a substantial reduction in thermal efficiency. Therefore, accommodating fluctuations within this timescale will involve operating plant on part-load, and this is again relatively inefficient. To the extent that the output from wave energy can be forecast accurately a number of hours ahead, the amount of plant that would have to be on part-load at any time could be kept to a minimum. Making some very broad assumptions about wave energy system characteristics over minutes to hours, it is possible that where wave energy represents only a few percentage points of installed system capacity it would have no more than a small effect on system control requirements and consequently the economic penalty would be small. As the installed capacity increases above about 5 per cent, the economic penalty associated with the greater demands for flexibility in other plant would increase significantly. Above about 20 per cent of capacity, wave power variability could be more of a problem, requiring compensating plant which is highly flexible and has a fast response. It is possible that this capability might be best met from pumped hydro or other storage plant of similar characteristics.

7.3 Economic assessment of wave energy

In assessing the economics of wave energy schemes the basic approach taken has been to quantify the benefits which wave plant can contribute to a generating system by way of fuel savings and possibly some firm power, and then to calculate a target capital cost (measured in terms of £/kW of rated output) at which a wave energy plant becomes an economic investment.

Preliminary calculations to determine the target capital cost for wave power have been made using the assumptions presented in Appendix 4. Because of a lack of data and information, particularly on certain aspects of wave energy, some of the assumptions had to be 'best guesses'.

From some of the earlier discussion it is evident that the interactions and interdependencies between generating plants means that the decision on whether a particular plant will be economic has to be made by examining the effects that it would have on the system as a whole. Making economic assessments in this manner is a complex subject: it would be inappropriate to attempt a full coverage of the subject in this paper but a simplified view is presented for clarity.

The basic costs of any power generation schemes are usually divided as follows:

(1) The capital costs which includes construction, interest charged on the capital expenditure incurred before generation commences and other incidental capital costs. It is also usual to include the cost of strengthening the grid transmission system.
(2) Operating costs such as maintenance and repair charges and fuel costs (in the case of wave power fuel costs are zero). Current estimates suggest that maintenance and repair charges for wave power systems could be significant, ranging from 2 per cent–5 per cent per annum of initial plant capital costs. There is, however, very little evidence to support any one particular estimate at this stage.

The benefits offered by a power generation scheme are again divided into two parts:

(1) First, a scheme may contribute firm power to the system. The firm power saves capital which would otherwise have to be spent on some other plant in order to maintain the required level of security of supply.
(2) Second, the operation of a new plant in the system enables the output from some other plants which have higher operating costs to be reduced. This gives a saving to the overall operation of the generating system which can be identified mainly as a saving of fuel. The savings are usually assessed over the life of the plant as it operates in merit order. The merit order of plant is determined by maximising the value of the savings less operating costs within the constraint of ensuring that demands are met.

Future generating systems
Since the economics of wave energy are to be assessed in the context of other plant in the system it is essential to make some assumption about the balance between the different types of plant in the system at a time in the future when wave power schemes of a significant capacity might be commissioned. Given a successful research, development and demonstration (R, D and D) programme the installation of wave power schemes could be underway around the turn of the century, or early next century. The analyses have therefore been carried out with respect to systems that might exist at that time. In particular, two important situations which differ in the availability of nuclear power are considered:

—in the first case it is assumed that nuclear power is limited and, although there would be some nuclear plant on the system, at the time when wave energy is an option it is not possible to construct as much nuclear plant as is economic;
—in the second case nuclear is not limited and there is no restriction on its selection.

In both cases it is assumed that any nuclear plant in the system operates at base load and therefore any increment of wave power plant would not reduce the load factor in any nuclear station, but would make savings by reducing fossil fuel consumption. Constraining future investments in nuclear plant so that it is always on base load could result in nuclear plant supplying less electricity than is economically optimum in the long term, but has the advantage of giving wave energy the best chance of being competitive, whilst providing a reasonable estimate for the capital cost targets of interest.

Since savings of fossil fuel will be the principal benefit from wave energy, and a major benefit from nuclear power, estimates of future fossil fuel prices will be important in determining the breakdown capital costs. Therefore, the analyses have been carried out for a range of possible fossil fuel prices that might occur around the turn of the century. It has further been assumed that

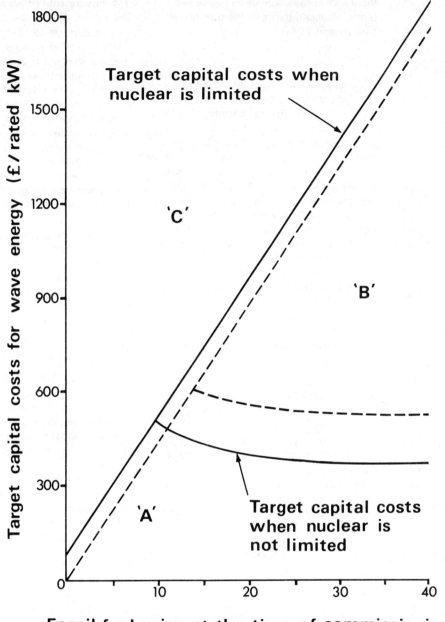

Fig 7.6 *Target capital costs for wave energy to be economic*
Notes:
1. The dashed line indicates the effect on breakeven capital costs of assuming that firm power has zero value.
2. Calculations have been carried out on the basis of a 5 per cent discount rate but this should not be interpreted as having any special significance. A 10 per cent discount rate would reduce the target cost by about a third in the case where nuclear is limited but has a much smaller effect in the case where nuclear is not limited.
3. The calculations assume that fossil fuel prices double over a period of 50 years from the time of commissioning (around the turn of the century).
4. For other assumptions see Appendix 4.

fossil fuel prices continue to rise in real terms, doubling between the turn of the century and 2050.

A limited-nuclear future. In the first situation, where nuclear power is a limited choice, wave energy will become economic when its capital charges are less than the net system operating savings, discounted over the life of the plant, plus any credit for firm power. For the purpose of these calculations a discount rate of 5 per cent has been used but this should not be interpreted in having any special significance. Figure 7.6 shows that the target capital costs rise with increasing fossil fuel prices but for a range of more probable fossil costs around the turn of the century it would lie between £1000–1500/kW of output rating. As a sensitivity it is useful to note that increasing the discount rate to 10 per cent would reduce the target capital costs by about one-third.

A future where nuclear is not limited. In the second situation nuclear is an allowable choice of plant and for most future fossil prices of interest it will be a competitive investment option. For wave power to be competitive under these circumstances it would have to be as good an investment as nuclear plant. As might be expected the target costs are more onerous in this case for most future fossil prices. Figure 7.6 suggests that capital costs might have to lie within the range £400–£500/kW of output rating. This result is relatively insensitive to the choice of discount rate.

Value of firm power

As a central assumption the value of firm power has been related to the cost of installing coal-fired plant. Taking a zero value for firm power indicates how, in the limit, the breakeven costs might change if very cheap bulk storage were to become available or if security of supply were to become considerably less important. It is noticeable that even in the limiting case of zero valued firm power the general conclusions for the wave energy cost targets are not substantially altered.

7.4 Future role of wave energy

The future role of wave energy is linked inextricably with its competitive position in relation to other generating plant and with the development of the markets for electricity. From the previous section it is evident that the competitive position could be influenced by whether nuclear power is a limited choice or not. In addition, the availability of nuclear power (or a competitive renewable energy source) is expected to have a significant effect on electricity's market. Thus, in a future where fossil prices rise substantially in real terms, the relative stability that nuclear power could bring to the cost of generating electricity is expected to improve the latter's competitive position, leading to faster rates of growth in demand for electricity than other fuels. If a restriction on nuclear plant resulted in more fossil fuels being burned to generate electricity, then electricity could be less competitive and therefore some lower rate of growth of electricity markets might be anticipated. The effect of these two general cases on electricity's future growth are illustrated in Fig 7.7. This idea can be linked with the analysis summarised in Fig 7.6 to derive some qualitative implications for the future role of wave power.

If the capital costs were to lie in range 'A' of Fig 7.6, so that wave power were competitive with nuclear, the arguments pertinent to the growth in nuclear electricity are largely relevant to wave energy. In these circumstances a successful R, D and D programme and adequate and timely investments could provide an alternative to some nuclear power or an insurance against a future nuclear disappointment to the extent that the potential for wave energy is adequate. Because the electricity generated by wave energy would be a highly competitive fuel there might be opportunities for sales into markets such as space and water heating where user storage could be installed to match the characteristics of the supply. These sales would be analogues to the off-peak electricity that is expected to be available from nuclear plant in the future.

If capital costs lie in range 'B' of Fig 7.6, then wave power would be confined to replacing the electricity that would otherwise have been generated from fossil fuels, as illustrated in Fig 7.7. Therefore the role of wave power would be more limited and would be particularly dependent on how

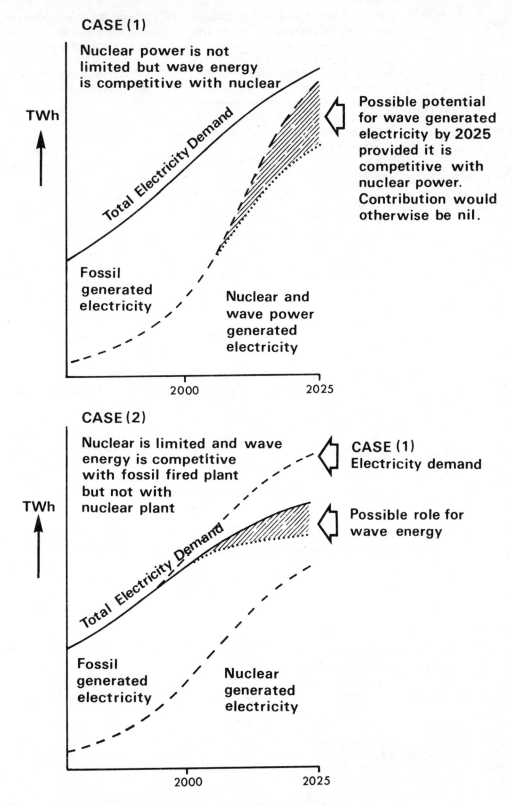

Fig 7.7 *An illustration of the possible roles for wave energy*

competitive fossil-generated electricity could be in energy markets of the future. The potential for selling wave power into the discounted price off-peak or interruptible markets, such as space and water heating, might be more limited in this case because of its higher cost. It could be more economic to ensure that the variability introduced by wave energy is fully compensated by alternative plant so that the output can be sold into the more valuable undiscounted markets.

Finally, wave energy capital costs might lie in range 'C' of Fig 7.6. In this region there is no cost argument for exploiting wave energy, provided only that sufficient fossil fuels are available to meet the postulated demand. If they were not, fossil fuel prices would rise even further and region 'B' would be expanded.

Preliminary cost estimates for wave energy power stations

As noted in earlier chapters the consulting engineers, in collaboration with the Device Teams, have prepared preliminary reference designs of those converter concepts for which appropriate data exist and have arrived at tentative costs. The reference designs have been sized at wave power stations with an installed generating capacity of 2000 MW. The results show that in the light of present knowledge and the early stage of engineering development, the predicted capital costs lie in region 'C' of Fig 7.6: wave energy, on these designs, would be uncompetitive with either nuclear or fossil generating stations in the future.

Preliminary attempts at costing can serve three valuable purposes:

— to give guidance in the formulation of energy policy:
  ● wave power cannot yet be included as a firm element in policy;
— to assist the choice between the various design concepts:
  ● the costing calculations are not yet sufficiently well founded to permit a clear choice;
— to highlight areas of R and D to which priority attention should be given in order to establish beyond doubt whether or not wave energy can ever be competitive with the other means of producing electricity.

In this third respect considerable progress has been made. A breakdown of the predicted costs in terms of percentage of the total is shown in Table 7.1. It is apparent that the critical cost centres are different for each design.

Whilst it is disappointing that at present the predicted costs suggest that wave power would not be economic, it must be pointed out that the figures are still preliminary and based upon a range of assumptions that have yet to be substantiated through the accumulation of appropriate

Table 7.1 Distribution of capital costs for various converter designs (all figures are percentages for each converter) (2000 MW installed capacity)

	Rectifier	Oscillating Water Column (NEL)	Raft	Duck	Flexible Bag
Construct concrete units	41	35	16	29	25
Structural steel	5	1	25	12	1
Mechanical components to power take-off	9	2	16	34	18
Turbines and electrical	22	11	12	3	11
Tow to site	10	3	3	2	3
Moorings	—	31	11	4	7
Power collection and transmission	3	7	7	6	23
Sundries and contingencies	10	10	10	10	12
	100	100	100	100	100

design data. In addition, there are a number of uncertainties within the assumptions used to establish the various target capital costs in Fig 7.6, although in many cases reducing these uncertainties is more likely to increase the stringency of the cost targets rather than relax them. There are a number of potentially optimistic avenues to be explored, which may well improve the competitive outlook for wave power, and the main value of the preliminary costing exercise has been to point the way into the immediate future of the R and D programme.

The size of the potential contribution of wave power is sufficiently large that at this stage it is worth considering as an insurance against the consequences of failure of one of our existing supplies.

Finally it should be noted that although wave energy converters would be massive structures and a substantial installation programme would create a large demand for the main constructional materials of cement and steel (greater than the rate required to build the largest North Sea oil platforms, and over a more prolonged period) the demands for these materials would be within the present capacity of industry and major new investments in them would not be required for wave power purposes.

8 Environmental and social aspects

Any human activity causes some disturbance to the environment and this may spread beyond the site directly affected. Impacts vary from very trivial or localised events, which may be scarcely detected, to major ecological change resulting in the replacement of one type of ecosystem by another. Whether the effects are considered to be beneficial or deleterious, and whether the latter are tolerable (taking all factors into account) is a matter for careful debate and judgement in each individual case. Moreover, there can often be competition between various forms of commercial activity which depend upon different features of the same environment. The technology of wave power has not yet reached such a state of detailed knowledge to allow firm conclusions to be drawn. However, the possible environmental interactions have been analysed in general terms during the two-year programme so as to provide guidance on more specific areas of knowledge which will need to be investigated before the installation of wave power stations on a large scale is commissioned. Because we are concerned with natural phenomena, some of these investigations may take many years to complete. However, in the meantime, no environmental effect has emerged which would raise major doubts about the ultimate acceptability of floating wave generating stations.

In addition to the environmental effects, some attention has been paid also to some possible social consequences. Because some of the most favourable wave sites in the UK are situated adjacent to sparsely populated areas, the social impact of the local availability of large amounts of competitively-priced power could be profound.

A Technical Advisory Group of the Wave Energy Steering Committee has been carrying out a preliminary examination of these matters (Appendix 3). A full environmental impact analysis will be possible only when a specific design of wave power generating station and specific installation sites have been chosen. On the other hand, it is not very helpful to deal entirely in generalities. The Technical Advisory Group decided, therefore, to devote most of its attention initially to the area of potential locations off the Outer Hebrides (Fig 8.1). If wave power is to make a substantial contribution to UK energy supplies, many generating stations must be located in this area since it has the highest wave energy availability. Therefore a detailed study of the area will be needed eventually, but in the meantime a general survey can show up the principles involved which could then be applied elsewhere if it were decided that a prototype or commercial demonstration plant should be built in one of the other possible areas shown in Fig 2.15.

Table 8.1 illustrates in matrix form the major interactions which will require analysis in due course. So far, attention has been directed in most detail at the effects of the wave energy converters themselves. The associated shore installations and power transmission lines, and shore-based activities concerned with construction and maintenance, will also be important but raise no new matters of principle which have not been faced in dealing with other types of power station.

The conclusions from the work done so far are presented below in four main sections:

—the possible effects of the converters on the local shore-line, through changes in the wave climate;
—the possible effects on fisheries, some of which may arise through changes in the tidal regime;
—the navigation of shipping;
—the social/economic development of the local communities.

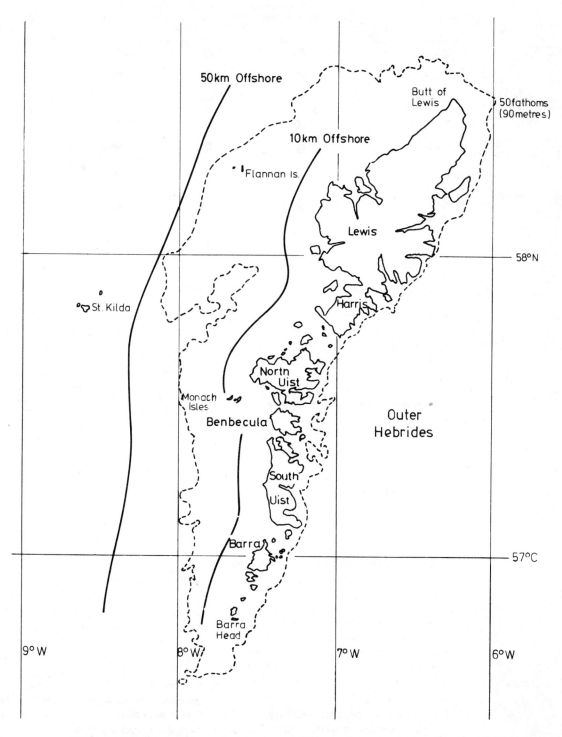

Fig 8.1 *Principal area studied in the preliminary review: between the 10–50 km lines off the Outer Hebrides*

Table 8.1 Interactions of wave power installations with other natural and human activities

	Wave power converter	Power transmission to shore	Shore installation
Marine ecosystems	●	●	●
Coastal ecological, geological and physiographic features	●	●	●
Navigation of ships	●	●	
Fisheries	●	●	●
Scenic beauty and tourism			●
Other local industries	●	●	●
Other national activities (e.g. defence)	●	●	

The dots indicate the major interactions which require analysis

Some important reverse effects—the influence of the marine environment on the wave energy converters—were discussed in Chapter 4.

8.1 Wave climate and the shore line

The very nature of wave energy converters is to extract some of the energy from the waves over a part of the total wave spectrum: exactly how much and the band width over which they will operate will depend upon the particular design adopted. The designer of the converter is concerned primarily with the amount of energy which it will extract, the environmentalist will be concerned with the amount which continues to be transmitted by the waves because it is that which may influence the physiography and ecology of the neighbouring shore-line.

A first analysis of the effect of converters on the wave climate has been made at the Hydraulics Research Station (HRS). The analysis has to compound a number of separate effects (see, for instance, Fig. 8.2):

—the height, direction and frequency of waves transmitted beyond the converters towards the shore;
—diffraction effects for waves filtering through the gaps between converters;
—re-injection of energy into the wave spectrum from the wind blowing between the converters and the shore;
—the bathymetry of the area.

Mathematical models do not exist which can combine all these effects. For instance, the present theories used in the prediction of waves from wind data refer to an initial state where the sea surface is either calm or has a natural spectrum created by wind action—and not to an artificial spectrum such as would exist in the lee of the converters.

However, simplified models were used at HRS in order to determine the nature and size of the overall problem. As an example one model predicted that, for converters assumed to be 20 m in depth operating in a storm characterised by a wind of 15 m/s (Force 7) blowing over a distance of 400 km or for 24 hours, the waves would be changed from a significant height of 5.9 m and a mean down-crossing period of 8.4 s to 4.6 m and 9.0 s respectively. The mean period is lengthened because the short-period waves would be reflected by the converters rather than being partially absorbed and partially transmitted.

In summary, it seems that the most noticeable effects of the converters on the wave climate at the shoreline will occur

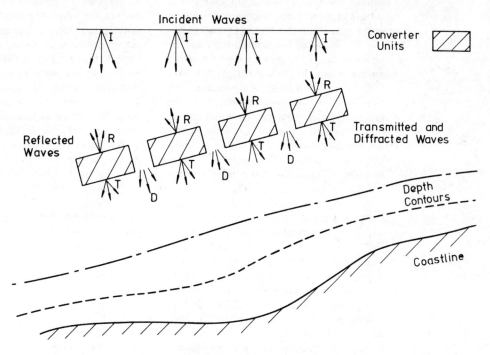

Fig 8.2 *Schematic view of a converter array*

during periods of medium wave activity where the combination of high reflection and high conversion in the converters gives significantly lower energy transmitted waves and wind action can only partly restore the energy in the distance between the converters and the shore. For major storms generated over longer distances or times the converters will have less effect since they are less efficient at absorbing waves of long period.

In the context of this chapter, the influence of the converters on the wave climate is important because of the effect of the latter on the beaches along the neighbouring shoreline.

Most attention so far has been given to the western coasts of South Uist, Benbecula and North Uist, since they are characterised by beaches of sand or shingle, which are the types most likely to be influenced by changes in the wave climate if such changes occur at all. Those in the area concerned appear to be in a state near to dynamic equilibrium, the changes from year to year in beach position, orientation, and slope being small. But the equilibrium over yearly cycles involves movement of sand in summer up the beach towards high water line, where it then dries and is blown by winds on to the dunes or 'machair' hinterland, and movement in the opposite direction in winter.

The small but significant effects of the wave energy converters on the wave climate may reduce somewhat the difference between the summer and winter beach profiles. In addition, since the steepness of the waves could be reduced if the converters are located less than 30 km offshore, there might be a net tendency for the beaches to accrete: the winter storms may well tend to cause less erosion than at present.

The limited amount of accretion which might take place is likely to be a benefit because it could increase somewhat the supply of wind-blown shell sand to the machair land, thereby tending to increase its extent and usefulness.

In summary, therefore, the effects of converters located off the particular coasts considered are more likely to be beneficial than detrimental to the beaches. The same conclusion would not necessarily be true for the effects of numbers of converters located off, say, the coasts of Cornwall or the Moray Firth. Each location will need to be considered in detail at a later stage if a serious intention emerges to locate the converters at a particular geographical site, but in the meantime this first examination of the problems has highlighted the basic principles involved and has forecast no detrimental effects in the Outer Hebrides area.

One final point must be mentioned, however. In the wave energy development programme the concept has emerged of mounting some types of converter, such as the Rectifier, fixed to the seabed and relatively close inshore. The environmental consequences could then become much more substantial but would be very dependent on local conditions and geography. This could become a major factor in deciding whether seabed-mounted devices are in fact feasible. A preliminary survey by the Consulting Engineers to the Wave Energy Steering Committee has indicated that the coasts of Benbecula and North and South Uist are more suitable than those of Lewis and Harris for fixed converters, from the viewpoint of the state and orientation of the seabed. This survey also revealed that a point of vital significance to the feasibility of fixed converters in these locations could be the fact that out to about 25 m depth there are dense forests of the seaweed Laminaria hyperborea. This grows to a height of two metres, or more, and the stipes, which are tough and fibrous, can achieve a diameter in excess of 8 cm. The fronds grow to a similar height and are cast annually. Fouling of seabed-mounted converters by such material could be serious, and substantial additional costs may be incurred by the need to introduce special anti-fouling devices or to construct the converters beyond the 25 m depth contour.

8.2 Possible interactions with fishery activities

The importance and size of the fishing industry in the UK render it inevitable that one of the major considerations in the deployment of wave energy converters must be the possible effect on fisheries. An initial survey has shown that the two fish to which most attention should be directed in the area west of the Outer Hebrides are herring and salmon.

Herring

In general, the areas of the UK coastal waters which are the most promising for wave power tend to overlap the areas which produce the largest catches of pelagic fish (that is, those which spend the major part of their lives in the surface or mid-water layers). Of the latter, the herring is the most important around the Outer Hebrides. These waters are believed from plankton and other surveys to contain some of the major UK spawning grounds of the herring at the present time.

The spawning takes place on areas of clean gravel, which have not been located precisely and may indeed change from season to season. Soon after the spawn hatches, the larvae (a few mm long) rise off the sea bottom and, having virtually no propulsive power, are carried along by currents. These currents can carry the larvae around the north of Scotland into the North Sea, where they may be caught as juveniles. Finally, as the fish approach maturity they migrate back to the spawning grounds where they form the basis of an important international fishery.

It is necessary to consider the interactions of a line of wave energy converters with a number of different stages in this life cycle. Because the converters would most probably be aligned roughly parallel to the seabed contours and would be of relatively shallow draft, and the main tidal propagation will also lie generally along the contours, it is unlikely that floating units beyond, say, 10 km from the shore will have any significant effect on the length or duration of the migration journey of the herring to the spawning ground. However, it might be possible for the converters to produce a small effect on the strength of the residual drift current.

Since the larvae are spread throughout most of the water column, their transport will be unpredictable in detail and will be affected significantly by wind and weather-generated currents. Whether the presence of wave energy converters could impose a noticeable overall effect on the interplay of all these natural forces is not known, but if they did then the implication would be that the larvae after a given elapsed time might find themselves a few tens of km away from the area in which they would otherwise have been, on average. Whether this matters, in terms of continuing food supply to the larvae, is also unknown—it may be either beneficial or deleterious or might indeed vary with locality.

On balance, in the very limited current state of knowledge, it is believed that the effects of wave energy converters on the drift of the herring larvae will be unlikely to have a significant influence on overall survival (natural seasonal variations in the survival of herring larvae are quite large in any case, and their causes are unknown). However, further information will need to be sought before the installation of converters on a large scale on:

— the drift currents;
— the behaviour of herring larvae.

More detailed knowledge of the drift currents will also be required in connection with the problems of mooring (see Chapter 5).

Perhaps the most important factor to emerge so far is that, since the herring spawn on the gravel areas of the seabed in this locality, major disturbance or removal of the gravel for construction purposes would appear to be inadvisable.

Salmon

Salmon migrate over long distances in moving between their feeding grounds in the ocean and their native rivers where they spawn. Salmon stocks are an important commercial and recreational resource: the value of the Scottish salmon fishery was estimated at over £20M in 1976. The migration routes are not known in detail, but it is possible to infer that the area immediately to the west of the Outer Hebrides may intersect some of the routes.

The pattern of movement is not simple. Fish tagged at Scottish sites have been found subsequently to the north, south, east and west of the point of release. However, there is a general impression that considerable numbers of salmon may move from the feeding grounds, for instance off Greenland and the Faeroes, towards the NW coast and thence disperse to the major Scottish rivers (including those on the East Coast) as well as those of Ireland and Wales.

In the locations concerned, the salmon probably swim near the surface, and the problem arises as to whether a line of wave energy devices could deflect them off course or whether the fish would simply swim through the gaps. We do not know enough about the salmon to be certain that they will be unaffected by a line of moving artificial objects. Moreover, the converters might create an environment favourable to large colonies of predatory birds, which would feed on the smolts, and of seals which might feed on the adult fish. The rapidly expanding colony of seals on the Monach Isles might recognise the converters as an advantageous additional habitat.

As for the herring, whilst these possible interactions must be considered further investigation may well show that a serious problem does not exist. In addition, a major installation of wave energy converters several hundred kilometres long could not take place in a short time but would be spread over many years. Experience gained with the early modules would be available in good time to take corrective action if that proved to be necessary.

Demersal fish and shellfish

Floating converters well offshore are considered unlikely to have a significant effect on demersal (bottom-feeding) fish such as cod, haddock, saithe and Norway pout. Similarly, no direct effect can be foreseen on the lobster population which has increasing commercial importance off the Outer Hebrides.

Converters fixed to the seabed and constructed close inshore may have much more significant effects, which might be both direct and indirect, and might be either advantageous or disadvantageous. The direct effect could be simply through the changed availability of the lobster habitats on the sea bottom, whereas the indirect effects could include increased hazards to the lobster fishing boats and pots (a benefit to the lobster but a disadvantage to the associated human commercial activities!).

8.3 Navigation of ships

The deployment of wave energy converters will present a hazard to shipping, exacerbated by their form—probably a very low free board which will render them relatively invisible to ships either by sight or by radar under most sea states.

Except for the English Channel, detailed

information on the pattern of shipping movements around the UK coasts is sparse. However, off the Outer Hebrides the present shipping movements are mainly concerned with the fisheries. Apart from the necessity to consider the influence of the deployment of arrays of converters on fishing operations generally, the hazards to fishing vessels can be of two kinds:

— collision, as noted above for all vessels;
— the formation of rougher seas, of particular importance to smaller boats.

It will be necessary to mark the positions of the converters with warning lights and radar reflectors. This will not be a trivial problem, since the markers may need to be sited at a greater distance above sea-level than the structure of the converters and will need to be relatively stable in the horizontal plane. Gaps in the line of converters, perhaps 2 km wide, would probably have to be left as navigation channels for fishing vessels and could well have to be marked to a higher standard than the converters themselves (eg there might need to be a light vessel or light and navigation buoy at each side of each navigational opening).

The requirements may turn out to be quite complex, but no new matters of principle appear to be involved and the matter will be considered by the appropriate responsible authorities as the occasion demands.

The hazards to small boats of rougher seas arises from the fact that part of the wave spectrum will be reflected by the converters. The reflected waves will combine in front of the converters with the incident waves to produce a complex pattern in which there will be increased wave heights, standing waves and an increased incidence of breaking waves. This pattern could be quite hazardous to small boats. However, it is possible that this effect will be confined to a distance close to the converters and the declaration for other purposes of an exclusion area around the converters may be sufficient to cover this point. Moreover, there would be a counterbalancing advantage from the fisheries point of view of calmer water behind the converters, though the exclusion area may have to cover this too.

Past experience has shown that an exclusion area would not entirely eliminate the consequences of the hazard or the chance of a collision, since human nature ensures that commercial attractions (prospects of higher fish catches) sometimes override prudence. The treatment as a statistical problem of the incidence of collisions between ships and fixed installations is receiving attention in the context of the North Sea, and no new work appears to be needed specifically for wave energy converters at present. However, the converters do introduce an aspect which will be very important to all navigational considerations: present concepts of mooring and mode of operation indicate that the converters cannot necessarily be regarded as fixed objects. Exclusion areas may have to be quite extensive to accommodate both the movement of the converters about a fixed point, if that is allowed in the design, and the positions of the moorings.

A new feature of the Outer Hebrides area which may emerge is the possible designation of clearways for the passage of deep draught oil tankers to and from the Sullom Voe terminal in the Shetland Isles. One of the clearways could pass between the Flannan Isles and the Isle of Lewis (see Fig 8.1), and the possible interactions with the precise location of some of the converters (especially off Lewis) would have to be resolved.

As noted at the beginning of this chapter, initial attention has been focused on the area off the Outer Hebrides. Preliminary information from other potential wave energy locations (see Fig 2.15) has not revealed any additional matters of principle but has indicated that problems concerning the navigation of shipping would be more complex than those in the Hebrides area. For instance, some possible locations off the Scilly Isles may intersect international waterways which could not be altered without international agreement, and if the converters were within 20 km of the southern Cornish coast careful analysis of the interaction with the important mackerel fisheries interests would be required.

A possible line for converters across the Moray Firth is indicated in Fig. 2.15. The complex nature of the potential interactions with fisheries operations is illustrated by Fig 8.3. These may be compounded with future exploration and production activities associated with offshore oil and gas.

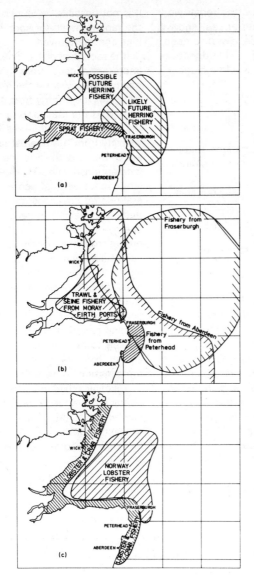

Fig 8.3 *Fishing activities in the vicinity of the Moray Firth*
(a) Pelagic fish (herring, sprat)
(b) Demersal fish (cod, haddock, whiting, plaice, lemon sole)
(c) Shellfish (lobster, crab, Norway lobster)

8.4 Economic and social development of local communities

As with the environmental issues, first attention in the preliminary review has been given to the Outer Hebrides area. The latter has no large scale industries and is characterised by a declining population and relatively high unemployment. The Western Isles Island Council has stated that its primary objective is the reversal of the trend of population decline due to the persistent selective out-migration from the area, and the improvement in the situation of the population in terms of employment. Installation on a large scale of wave energy converters could assist the attainment of this objective in two ways:

— the labour force needed to operate and maintain the system;
— the possibility of utilising some of the power to establish new industries in the area.

A first survey of these issues has been carried out by the Highland and Islands Development Board (HIDB). Whilst the prospects for setting up major industrial developments in a remote and difficult area with a sparse population may appear daunting, there are a number of examples where this has been achieved successfully, for example the Dounreay Nuclear Establishment, the Uist Rocket Range, the Kishorn construction site, Sullom Voe oil terminal, and oil-related developments on Flotta. Whilst introducing the industrial infrastructure associated with wave energy would be a formidable task, the evidence from these earlier activities indicates that the difficulties would be by no means insuperable. The HIDB study also indicated that the availability of land suitable for such development should not be a constraint.

Construction of the converters may well take place elsewhere because of the very high tonnage involved, the converters then being towed on to station. However, a small fleet of tugs and supply boats would be required which could be based locally. Whilst the main ports in the Outer Hebrides have developed historically on the eastern coast (ie looking away from the Atlantic), and the passages between the Isles virtually all present difficulties, there are some suitable locations on the western coasts.

The inspection and maintenance of moorings for the converters will be vital—the converters cannot be allowed to break adrift—and a small workforce of skilled divers would be needed, although probably less than for the offshore oil industry in the North Sea.

A much larger requirement than the supply bases for equipment and skilled manpower could arise from the need for one or more maintenance bases. There is likely to be a variety of maintenance and repair work which cannot be carried out at sea, and will require the converters to be towed to deep well-sheltered water with possible dry-docking. Criteria for such sites would include:

—shelter from sea and swell;
—adequate depth of water;
—entrances and configurations suitable for manoeuvring the unusual shapes of the converters;
—land at the shore physically suitable for development;
—road access or the capability of road access;
—minimum practicable towing distance from the sea stations.

The survey showed that it would be difficult to identify any particular sea loch which would meet all these criteria. It was pointed out that the labour force required to operate a maintenance base for a large array of converter stations might run into thousands of men rather than hundreds, much of it highly skilled. Until both the design of the converters and the logistics of the whole system have been developed much further, it is not possible to be more precise about the requirements for such shore-based facilities and manpower.

It is recognised that one of the most important aspects from the visual amenity point of view will be the method and route by which the energy from the converters is carried to the mainland (it is unlikely that the converters themselves will intrude on visibility from the nearest shore, unless they are of the type fixed to the seabed and located close inshore). However, the problem cannot be analysed in detail until several related decisions have been clarified. In particular, decisions between electricity and energy-intensive chemicals as the desired output from the system, and between local use of electricity for new industries or feeding to the national grid, see Fig 8.4. The most difficult of the visual amenity problems is likely to arise from the fact that a system consisting of a large array of converters feeding electricity into the national grid is unlikely to be able to avoid transmission across Skye. Very detailed and careful transmission route planning will be needed.

A similar preliminary review of other potential wave energy areas has not yet been carried out since, based on the data on the availability of the wave energy, they would be less likely to attract the first very large

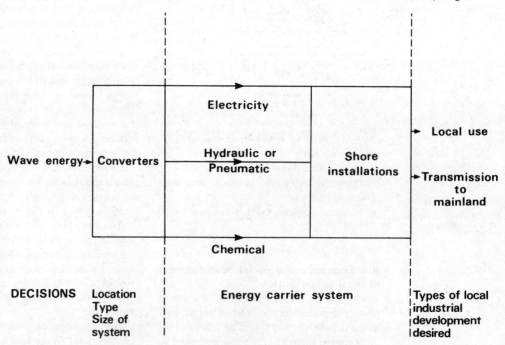

Fig 8.4 *Systems decisions needed in order to carry out a detailed examination of the amenity consequences of wave energy for the Outer Hebrides*

arrays of converters which will be needed if wave energy is to make a substantial contribution to the UK's long-term supply.

8.5 Concluding remarks

A first broad examination has been made of the environmental and local economic development consequences of the installation of large arrays of wave energy converters off the Outer Hebrides—the most suitable location from the viewpoint of the availability of the wave energy. No major deleterious environmental effects of the converters themselves could be identified, provided that the converters are well offshore, but more detailed information should be sought on aspects relating to herring and salmon fisheries so as to confirm that they will not be affected significantly. It would be inadvisable to base construction techniques on the extraction of gravel on a large scale from the seabed in this locality because of the importance of the gravel deposits as herring spawning grounds. Further studies are planned to examine the nature conservation implications of the proposed schemes.

The advent of economic wave-derived energy would present many new opportunities for the development of both the traditional seafaring and new industries as a substantial contribution to local economic wellbeing. Schemes for feeding wave energy as electricity into the national grid will probably involve transmission across Skye and this would give rise to considerable visual amenity problems.

Converters fixed to the seabed and located close inshore would involve more significant environmental consequences than floating converters further out to sea, and these will require much more detailed examination when the designs and locations have been decided. In addition, arrays of such converters could represent a **permanent** and possibly undesirable change to the environment unless some technically and economically acceptable method for their removal at the end of their service life can be devised. In a reverse sense, the environment itself could have a much more profound influence on seabed-mounted converters than on floating converters: particular attention will be needed to fouling by seaweed for locations off the Outer Hebrides.

Preliminary information on other areas of interest for wave energy, such as the Moray Firth and the SW coastal waters, indicates that the possible interactions with the navigation of shipping and with fishery activities will require detailed assessment in advance of the installation of large numbers of converters.

Appendix 1 Principal contractors in the Department of Energy programme (at September 1978)

Development of converters
Hydraulics Research Station, Wallingford
National Engineering Laboratory, East Kilbride
Sea Energy Associates Ltd, Cheltenham
University of Edinburgh
Vickers Ltd, Eastleigh
Wavepower Ltd, Southampton

Technical Advisory Groups—generic work

TAG 1
University of Lancaster

TAG 2
Flight Refuelling Ltd
Institute of Oceanographic Sciences
Marine Exploration Ltd
National Engineering Laboratory
National Maritime Institute

TAG 3
British Shipbuilders
Harwell, Materials Development Division
Lloyds Register of Shipping
National Engineering Laboratory
National Maritime Institute
Scottish Marine Biological Association
The British Ship Research Association

TAG 6
Dunlop Ltd
General Electric Company Ltd
International Research & Development Co. Ltd
Lucas Group Research Centre
Pirelli General Cable Works Ltd
University of Edinburgh

TAG 7
Hydraulics Research Station
Nature Conservancy Council

Consultants: Rendel, Palmer & Tritton, in association with Kennedy & Donkin

Appendix 2 Wave Energy Steering Committee (1978)

Terms of reference
1. To draw up and agree a national programme of work for the study of wave energy.
2. To advise on the implementation and management of that programme.
3. To advise on the technical briefing of UK delegates to international meetings on wave energy.
4. To report to the Chief Scientist, Department of Energy on matters relating to wave energy.

The present membership is:
 Dr F J P Clarke
   Harwell—Chairman
 Dr A M Adye
   Science Research Council
 Dr J K Dawson
   ETSU
 Mr G A Goodwin
   Department of Energy
 Mr D C Gore
   Department of Energy
 Mr R Hancock
   Department of Industry
 Mr R C H Russell
   Department of Environment
 Dr C S Smith
   Admiralty Marine Technology Establishment
 Mr D S Townend
   The British Petroleum Company Ltd
 Dr J K Wright
   Central Electricity Generating Board
 Mr P R Wyman
   General Electric Company Ltd
 Mr C O J Grove-Palmer
   ETSU—Secretary
 Mr L A W Bedford
   ETSU—Minutes Secretary
 Mr P Clark
   Rendel, Palmer & Tritton—Consultants to WESC

Appendix 3 Technical Advisory Groups

TAG 1—New designs
Terms of reference
1. To review the fundamentals of wave motion in relation to the principles of wave energy extraction and, where thought appropriate, either to encourage and/or support current fundamental research work or promote such work particularly on basic principles not covered by the current WESC or SRC programmes.
2. To review on behalf of WESC all ideas and applications for financial support for new designs of wave energy converters submitted to the Department of Energy, and to make recommendations for action.
3. To recommend and initiate work on generic concepts associated with the fundamental principles of wave energy converters including aspects of current designs, drawing on resources of expertise available from the wave power programme as a whole.
4. To liaise with SRC in respect of research at universities on wave energy devices.
5. To keep informed on new international developments and to assess their relevance to and implications for the UK programme.

The present membership is drawn from:
 The British Petroleum Company Ltd
 Central Electricity Generating Board
 Department of Energy
 University of Cambridge
 ETSU
 Rendel, Palmer & Tritton—Consultants

TAG 2—Wave data
Terms of reference
To determine what wave data are needed by the engineering development teams and what is required by other Advisory Groups, and to arrange for such data to be provided if it already exists. If there are requirements for additional data the group will propose programmes of work to acquire such data.

The present membership is drawn from:
 Central Electricity Generating Board
 Department of Energy
 Energy Technology Support Unit
 Hydraulics Research Station
 Institute of Oceanographic Sciences
 Marine Technology Support Unit
 Meteorological Office
 National Maritime Institute
 Lanchester Polytechnic ⎫
 Vickers Ltd. ⎪
 National Engineering ⎬ Device Teams
  Laboratory ⎪
 Wavepower Ltd. ⎪
 University of Edinburgh ⎭
 Rendel, Palmer & Tritton—Consultants

TAG 3—Structures and fluid loading
Terms of reference
1. Examine loading and structural design aspects of proposed converters, consulting with Device Teams as necessary.
2. Advise the Wave Energy Steering Committee, and where appropriate the Device Teams, on the problems of wave action and structural design.
3. Advise the Steering Committee on structural adequacy and efficiency of proposed converters, making recommendations where appropriate for improvements to structural design.
4. Evaluate programme of development work being carried out by the Device Teams, initiate and monitor on behalf of the Steering Committee any supplementary R and D which may be necessary to ensure the accurate prediction of wave-induced loads and mooring forces and efficient structural design.

The present membership is drawn from:

    Admiralty Marine Technology Establishment
    British Ship Research Association
    Cement and Concrete Research Association
    Central Electricity Generating Board
    Energy Technology Support Unit
    Harwell, Metallurgy Division
    Hydraulics Research Station
    Lloyds Register of Shipping
    National Engineering Laboratory
    National Maritime Institute
    Lanchester Polytechnic ⎫
    Vickers Ltd. ⎬ Device Teams
    Wavepower Ltd. ⎪
    University of Edinburgh ⎭
    Rendel, Palmer & Tritton—Consultants

TAG 4—Mooring and anchoring
Terms of reference

1. To advise on the suitability of present mooring and anchoring systems to the requirements of various wave power converters.
2. To advise Device Teams on technology specific to their designs, and to make recommendations on research and development requirements.
3. To recommend, instigate and monitor research and development programmes of a generic nature on behalf of the Wave Energy Steering Committee.

The present membership is drawn from:

    British Petroleum Ltd.
    Energy Technology Support Unit
    Harwell, Materials Development Division
    Hydraulics Research Station
    Lloyds Register of Shipping
    National Engineering Laboratory
    National Maritime Institute
    Offshore Supplies Office
    Pirelli General Cable Works Ltd.
    Vickers Offshore Ltd.
    Lanchester Polytechnic ⎫
    National Engineering Laboratory ⎬ Device Teams
    Wavepower Ltd. ⎪
    University of Edinburgh ⎭
    Rendel, Palmer & Tritton—Consultants

TAG 6—Generation and transmission
Terms of reference

1. To identify possible energy conversion, generation and transmission systems.
2. To estimate the performance and cost of the more promising systems and make a first order assessment of the impact of the operational and performance (transfer efficiency) characteristics of particular designs on the overall economics of converters.
3. To provide design information for the teams developing particular converters.
4. To estimate the timescales and the R and D effort required to implement particular designs.
5. To make recommendations to the Wave Energy Steering Committee on the most promising system(s) for development.

The present membership is drawn from:

Contractors group
    Central Electricity Generating Board
    Davy-Loewy Ltd
    Department of Energy
    General Electric Company
    International Research & Development Ltd
    Lucas Group Research Centre
    Pirelli Cables Ltd
    Kennedy & Donkin/Rendel, Palmer & Tritton—Consultants

Device team group
    Central Electricity Generating Board
    Energy Technology Support Unit
    Hydraulics Research Station
    National Engineering Laboratory
    Ready Mixed Concrete Ltd
    University of Edinburgh
    Vickers Ltd
    Wavepower Ltd

TAG 7—Environmental impact
Terms of reference

1. To examine the possible environmental effects of large wave power stations on:
(a) the morphology of the adjacent coastline;
(b) the navigation of shipping;
(c) the local ecological balance;
(d) the fishing industry;
(e) leisure activities;
(f) interaction with other activities, eg Ministry of Defence, oil exploration, etc.

2. Bearing in mind the likely location of wave power stations, to develop an awareness of the impact of shore installations on areas of scenic beauty and of the availability of large amounts of energy in areas of low economic activity.
3. To consider environmental effects in geographical areas which might be used for Phase II of the development programme.
4. To report to the Wave Energy Steering Committee and to advise the Device Teams as required.

The present membership is drawn from:

Department of Agriculture & Fisheries for Scotland
Department of Trade, Marine Division
Energy Technology Support Unit
Highlands & Islands Development Board
Hydraulics Research Station
Ministry of Agriculture, Fisheries & Food
Ministry of Defence
Nature Conservancy Council
Scottish Marine Biological Association

Appendix 4 Main assumptions used in calculating breakeven capital costs of wave energy

		Nuclear plant	Coal plant	Wave energy plant
1. Plant characteristics				
Average plant life	(years)	25	30	25
Mean equipment availability	(%)	70	70	85
Firmness at peak	(%)	85	85	25
Mean annual load factor	(%)	70	up to 70	45
2. Plant costs (1)				
Total capital costs (2)	£/kW	730	400	?
Operating costs (excluding fuel)				
—Fixed operating costs	£/kW pa	5	2	⎫ 3% plant
—Variable operating costs at an annual load factor of 100%	£/kW pa	5	20	⎭ capital pa
Net present worth lifetime fuel costs	£/kW	450	(3)	nil
Net present worth gross system operating savings	£/kW	1900 (4)	(4)	1280 (4)

Notes:
(1) All costs are quoted in real terms using 1978 money values. The costs are illustrative but are intended to be appropriate at the turn of the century. A discount rate of 5% is used throughout.
(2) Total capital costs include construction costs, interest during construction, transmission costs and initial fuel for nuclear plant.
(3) Coal plant fuel costs depend on the thermal efficiency of the plant and on the load factor assumptions. Coal plant is assumed to have a thermal efficiency of 35% which is typical for modern fossil plant and it is assumed that this type of plant will be the means of generating electricity from fossil fuels during the early part of next century.
(4) The gross system operating saving is evaluated at a fossil fuel price of 20p/therm. At any other assumed price it will be directly proportional to the value given. Gross system operating savings consist mainly of the value of the fuel saved when a new plant is operated in the system and in deriving Fig 7.6 only this saving is assessed. It is also assumed that all new plants which make a saving in this way displace fossil fuel and this implies that the available nuclear capacity is always less than the minimum demand. For coal plant the gross system operating savings are approximately equal to fuel costs.
(5) As a central assumption the value of firm power is based on coal plant and is taken as £430/kW. The effect of a zero value is also indicated in Fig 7.6.
(6) Other assumptions are discussed in the main text.